HOW
TO DO
MATHS
SO YOUR
CHILDREN
CAN TOO

For Yasmin – and for everything she gave us

HOW TO DO MATHS
SO YOUR CHILDREN CAN TOO

THE ESSENTIAL PARENTS' GUIDE

NAOMI SANI

Vermilion
LONDON

7 9 10 8 6

Published in 2010 by Vermilion, an imprint of Ebury Publishing

Ebury Publishing is a Random House Group company

The Random House Group Limited Reg. No. 954009

Addresses for companies within the Random House Group can be found at
www.randomhouse.co.uk

A CIP catalogue record for this book is available from the British Library

Designed and typeset by seagulls.net

ISBN 9780091929381

Copies are available at special rates for bulk orders. Contact the sales
development team on 020 7840 8487 for more information.

To buy books by your favourite authors and register for offers, visit
www.randomhouse.co.uk

The Random House Group Limited supports The Forest Stewardship
Council® (FSC®), the leading international forest-certification organisation.
Our books carrying the FSC label are printed on FSC®-certified paper.
FSC is the only forest-certification scheme supported by the leading
environmental organisations, including Greenpeace. Our
paper procurement policy can be found at
www.randomhouse.co.uk/environment

CONTENTS

ACKNOWLEDGEMENTS

My gratitude goes to everyone who has encouraged me to write this book. In particular Ruth Keily, who believed in this book from the beginning, and used her brilliant editorial guidance to show me how to write with confidence. Her expertise has been invaluable.

My thanks to all the children I have taught: you have enriched my life; and to all the teachers who have inspired me: you do an amazing job. Also to the teachers at Bathwick St Mary Primary School, Bath, for their time and interest.

Thanks to Professor David Burghes and the rest of the team at the Centre for Innovation in Mathematics Teaching at Plymouth University for their ever-enlightening research as well as their constant drive and determination to raise the standards of maths teaching.

I am grateful to my editor at Vermilion, Julia Kellaway, who championed this book and remained committed even when it was significantly delayed.

A big thank you to my gorgeous boys, Oliver and Matthew, for providing daily insight into how young children learn about numbers. And finally, thanks to my husband Phil, without whose constant support, undoubting belief and extra childcare duties this book could not have been written.

INTRODUCTION

Maths is an emotive subject. Some people love it; some people hate it. But everyone has to do it.

Almost everyone agrees it is important; but not everyone is sure why.

And a lot of parents would heave a sigh of relief if their children were able to 'just do it' – easily, effortlessly and successfully. The next Einstein may not be their ambition but a good GCSE grade certainly is.

Battle, difficult, struggle, effort, hard work, painful... for a lot of people, these are the words that spring to mind when talking about their own experience with maths at school. Invariably when people find out I am a maths teacher they comment on how clever I must be. I'm not. I just like maths and I was lucky enough to have some outstanding maths teaching from a very early age. And I believe that's the crux of the whole thing. If maths is explained clearly and simply, it is not only a very 'do-able' subject but also a very enjoyable one.

The way maths is taught in schools today has changed significantly over the last few years in an attempt to do just this and make maths more 'do-able' and more enjoyable. But as a result parents are often left struggling to recognise what is being taught and are undoubtedly unsure of how it is being taught.

So this is the *raison d'être* for this book – to explain things simply and to highlight the new methods, terminology and teaching ideas.

As a teacher, and a mum with a child in primary school, I have been asked countless times if there is a book I can recommend to help parents help their children with maths. There just hasn't been such a book – so I decided to write one. When I talked to other parents about my idea for this book, they have all been really encouraging and enthusiastic, but just a little disappointed that I hadn't written it already!

WHO IS THIS BOOK FOR?

This book is primarily aimed at parents (and here, as in the rest of the book, I include guardians, carers, grandparents and any other interested adults). It will be of particular relevance to parents of children between the ages of 5 and 12, but is likely to help parents of older children who need a bit of extra support.

The book is for parents who are numerate but are unsure of the new way of doing things as well as for parents who would like some support and help themselves.

However, this book could also be of help to trainee teachers, teachers without a maths specialism and classroom assistants.

WHAT IS THIS BOOK FOR?

This book sets out to do three things:

- Explain maths topics with a simple step-by-step approach.
- Explain the new way of doing things in the maths classroom.
- Empower parents to feel able to help their children with maths.

In my experience there is a greater demand for a book to help with maths than for any other subject – but why should this be? I think the reasons for this boil down to some very basic anxieties about maths:

- Some of us feel we are bad at maths ourselves.
- We know maths is vitally important to our children's future success.
- Many of the terms and methods used in the classroom have changed since we were at school.

PARENTS' OWN FEELINGS ABOUT MATHS

Many adults feel they were poor at maths themselves. In most cases this is simply not true. Maths teaching was often so poor, and expectations so low, that many pupils failed to reach their potential altogether. The number of pupils passing the old 'O' level was shockingly low, at around 25 per cent. That means only the top quarter of students managed to secure a grade 'C' or better. This figure should be seen for what it was – a reflection of a poor system, not poor students. And yet I know lots of adults who *do* have an 'O' level in maths and still feel they are useless at the subject. This simply doesn't add up!

Something is wrong if a person who achieved marks in the top 25 per cent still feels a failure. In my view, it's about how – as children – those adults were made to feel about maths. Often they lacked confidence in what they were doing, didn't enjoy the lessons and didn't see the point of the subject.

And if the top achievers felt that way back in those days, spare a thought for everyone else.

Don't pass it on!

So here we have thousands of adults, many of whom are now parents, who feel they are rubbish at maths. At this point, I'd like to assert a simple but hugely important rule:

> PLEASE, PLEASE… NEVER say that you are bad at maths – not anywhere within a 100-mile radius of any child you ever want to influence.

As a parent, you know you have incredible influence as a role model. What you think and say is what your children will think and say. If you say you are bad at maths – you are giving the very loud and clear message that it is okay for your children to be bad

3

at maths. And we all know that is just not true. Use your power wisely, and perhaps our generation will manage not to pass on the negative feelings many of us were given about maths when we were young.

You don't have to lie. You don't have to pretend to be a rocket scientist. You are allowed to say you wish you were better, or you'd like to learn more, but the truth is you're probably not as bad as you think you are. I hope this book helps too.

And this is really important for teachers too. Many excellent and otherwise outstanding teachers have been heard to say to pupils (and almost boast) that they are no good at maths. I think their intentions are honourable. I imagine they are seeking to reassure and help the children. However, their actions will in fact do the complete opposite. Instead of helping, they will be hindering. Instead of reassuring, they will be reinforcing negative ideas – that it's okay to be bad at maths. IT IS NOT OKAY.

WHY IS MATHS SO IMPORTANT, ANYWAY?

This is a really good question, but a hard one to answer succinctly.

Mathematics has a rich history and is one of the oldest academic subjects. Maths is about pattern, structure and order and there is no doubt that mankind has been on an endless quest to find structure and order in an attempt to answer every conceivable question.

For example, computer programming animation is dependent on mathematical know-how. In the first *Toy Story* film there is no water because the mathematics to simulate water simply didn't exist. A few years on and water and splash effects appeared in *Finding Nemo*. The maths needed to simulate water had been 'discovered'.

In today's world maths is everywhere around us. When driving a car we can judge from our speed and the distance we need to travel approximately how long our journey will take. In order to balance a bank account we need to know about debit and credit. To manage the mortgage we need an understanding of percentage interest rates. When planning a wedding, a family holiday or other special occasion we have to keep within a budget.

Maths is fundamental to modern science and technology. Without maths there would be no aeroplanes to fly you away on holiday, no ultrasound scan to let you see your unborn baby, no rockets to put satellites into space, or – and this *already* seems difficult to imagine – no internet for us to use. Life would be very different.

A surprisingly wide range of careers have maths at their core, including computer graphics, engineering, meteorology, medicine, retail and banking. Even less obvious career routes, such as law, politics and marketing, demand strong numeracy skills in order to rise to the top.

And without doubt there is an intrinsic beauty to maths. Studying maths for the aesthetic appeal alone is enough for some. There can be a definite buzz of excitement and satisfaction when a maths problem is solved.

I also believe studying maths is good for the brain: a form of mental gymnastics. It can stretch you, challenge you and give your brain some worthy exercise. It is true that for the brain, like muscle tone, if you don't use it, you lose it!

But is all this a convincing enough argument to commit our children to compulsory maths education until they are 16? It may well be true that no technological advances would be made without a high level of expertise in mathematics, but what if we are not expecting our progeny to be technological whizz-kids?

Just as everyone deserves to be literate, I believe everyone deserves to be numerate. Being literate can open minds and open eyes. Being numerate can do the same. Daily life is made easier; doors of opportunity are opened wider. In this fast-paced, technology-driven, ever-changing world in which we live – now is not the time to embrace ignorance.

One of the reasons that there have been significant changes in the classroom in recent years was the need to address the low levels of numeracy among a startlingly large number of children.

CHANGES IN THE CLASSROOM

So what exactly is going on in the classroom? Good things, in fact. Compared to our own school days, expectations for mathematics are much higher. Government interest, intervention and initiatives, coupled with improved methodology and higher standards of teaching, mean kids have never had it so good.

But this improved methodology has left many parents floundering. For most of us, it 'just doesn't look like it did in my day'. Even numerate parents can find it difficult to relate to the new ways children are being taught in class.

A highly intellectual scientist – my dad – was always offering to help me with my physics homework. Yet I used to end up in tears of frustration every time because he just didn't set it out like the teacher, and so I didn't understand it.

An outstanding primary teacher recently addressed a group of parents at my son's school. Talking about homework, and maths homework in particular, this teacher explained how crucial it was for parents to recognise how children are taught today and how children now work things out. If parents don't know then their children can get panicky, upset or even 'shut down' altogether when it comes to asking for help.

The new methods are sound, and most are really quite similar to what we would have learned in school. We just need to be able to recognise them and learn the terms that children will use. Then we can explain them to our children without the tantrums.

Understanding maths is about building blocks. Get it right from the start and you've built a tower to be proud of. Get it wrong and that future high-rise building is a bit wobbly at best. Without the right foundations, some children will fail to build a tower at all.

Armed with the information in this book, we can feel more able to help our children build on up from the first few blocks, starting with the simplest levels of numeracy and progressing towards what would be expected of pupils in their early years at secondary school.

HOW TO USE THIS BOOK

Chapter 1: 'The Number Tool Kit' introduces concepts and ideas common to the other chapters. So Chapter 1 is *the* place to start, after which feel free to dip in, skip over, jump ahead or revisit different parts of the book, as you need to.

As your children progress through school, this book enables you to stay one step ahead and be able to offer help on maths when it's needed. You will be able to reinforce classroom practice and be confident that you are not contradicting the methods currently used at school. There are step-by-step worked examples throughout each chapter.

Each chapter of this guide is divided into three sections:

- **Understanding the Basics** is roughly related to Reception class and Years 1 and 2 or 5–7 year olds.
- **Moving Forward** is roughly related to Years 3 and 4 or 7–9 year olds.
- **To the Top** is roughly related to Years 5 and 6 and onwards into secondary school or 9–12 year olds and older.

Understanding the Basics not only relates to the first few years at school (Reception class, Years 1 and 2) but is also of great help to older pupils needing to recap or consolidate their knowledge.

Moving Forward is written with Year 3 and Year 4 pupils in mind and will also be of help to older pupils needing support.

To the Top will help you if you have children in Years 5 and 6 or in the early years of secondary school.

How the school 'year' groups relate to ages and Key Stages is shown in a simple reference table in the Appendix at the back of the book (see page 372).

Children progress at different stages. It's not important whether your children 'should' be doing something by a certain age. Much more significant is how thoroughly they understand principles and techniques – with your help, if needed.

Please don't feel under any pressure to rush your children just because somebody else's child is apparently 'ahead' of yours. It will honestly do more harm than good, and your children may miss the opportunity to grasp the fundamentals fully.

Maths is a very broad and far-reaching subject. This book concentrates on the core numeracy topics, focusing on those points that parents may well have forgotten as well as on teaching methods that have changed considerably. Not all maths topics can be fully addressed in this one book. Three areas of the primary curriculum, Shape, Measuring, and Statistics (or 'Handling Data'), have not been dealt with in detail. Instead, useful terms, definitions, reminders and formulae are included in the Glossary (see page 357).

At this point, it's worth just touching on some general themes relevant throughout the book.

THINKING TIME

Allowing your children time to think before expecting/demanding an answer gives their brain time to compute. This may be more important in maths than in any other subject. It's not just a simple matter of recall. If you don't allow this crucial thinking time and rush to fill the silence with an answer you are denying both yourself and your children.

A child can feel robbed of the feeling of success that comes from working something out well. And you can feel frustrated that the child 'just isn't getting it'. Another side effect can be that children soon 'learn' not to bother trying to think for themselves.

You see this in maths classrooms all the time. Teachers, ever anxious to keep the pace and momentum of the lesson flowing, can be nervous of silence. They then either answer their own question or rely on the faithful few who are always quick and ready to put their hands up to please the teacher. In my experience a 'no-hands-up policy' works best as the whole class then has to be on 'standby'!

Waiting can actually be harder than it sounds. Something I always do to force myself to wait – with my own child as well as children I teach – is to count to 10 in my head. You may be pleasantly surprised at how often they *can* 'work out' the answer.

KINAESTHETIC LEARNING

Kinaesthetic learning sounds complicated but it simply means learning through touch and movement. Children, especially of primary age, like to move around. No surprises there for parents, but most children actually do learn best when physical movement is involved.

Perhaps if every lesson could be an all-singing, all-dancing affair we'd have no problem with pupil motivation, but I'd worry about legions of frazzled teachers limping away from an already depleted profession. Meanwhile, our keen kinaesthetic learners can be catered for quite easily.

For example, when we talk to children about adding, let them imagine leaping and bounding along the 'Number Line' (see page 19). Have them visualise hops, skips or jumps. Encourage them to imagine they are frogs-a-leaping or kangaroos-a-bouncing. Better still, let them jump a little themselves, at least in the very early years.

YOUR CHILD HAS STYLE... LEARNING STYLE

It is generally accepted that there are three different approaches – or 'styles' – of learning. Which of the descriptions below best describes your children?

Visual learners

Visual learners learn best through seeing. These learners may think in pictures and will learn best from visual displays like posters, illustrations, drawings or film footage. They also need to see the teacher's body language and facial expressions to fully appreciate what they are being taught.

Visual learners benefit from sitting at the front of a class to avoid any obstructions ahead of them. While listening to a lesson or lecture, visual learners may need to make detailed notes, pictures and doodles to help them 'see' what they are being told.

Auditory learners

Auditory learners learn best through listening. Talking things through, listening to others, discussions and debates all help these learners. Auditory learners pay close attention to tone of voice, pitch and pace to help fully appreciate what they are being taught. Written information might not mean much to them until it is read out loud.

Kinaesthetic learners

Kinaesthetic or tactile learners learn best through movement. A hands-on approach to learning suits them best. They need to 'touch' and 'do'. Kinaesthetic learners often find it hard to sit still and need to be offered opportunities to explore what they are being taught actively.

Ideally all three styles of learners should be catered for in all lessons. This is not always possible, but it is still worth knowing how your children prefer to learn. Armed with this insight, even a small change (like moving seats in the classroom) can make a big difference, and could help you support homework too. Being given a doodle-pad can help some children concentrate and focus on the content being delivered. Some secondary schools now issue these as standard.

Generally speaking, most young pupils, *especially boys*, enjoy kinaesthetic learning and you can often see this in primary schools. As pupils move to secondary school fewer opportunities for kinaesthetic learning are offered, as a general rule. This can be a problem, *particularly for boys*, and it may be an area where you, as a parent, can offer useful alternatives.

IT'S ALL IN THE MIND

In the first few years of primary school, we need to stick to 'mental methods' only. Parents are sometimes surprised to find that young children are expected to write down very little in their maths lessons. Even by Year 3, only very informal pen-and-paper jottings are used.

The reason for this is to enable children to build a strong understanding of numbers and appreciate and visualise the underlying ideas – before being burdened with algorithms (that is, standard written methods).

The ability to calculate mentally is essential for all – and efficient strategies are now taught explicitly. Mastering mental methods should never be left to chance!

As mental methods are refined and strengthened, they are used to support informal written methods. These methods become more efficient and succinct and then lead to more formal standard written methods.

The goal, towards which we should all strive, is that, by the end of Year 6, all children are equipped with a set of 'tools' (both mental *and* written methods) that they can use and understand correctly. No child should ever be calculator-dependent.

Only when they have a sound understanding should symbols (such as '+', '–', '×', '÷' and '=') or other mathematical language be introduced. As with everything concerning child development, some children will reach this stage earlier than others.

So, in the early years, your children may be super keen, but they are unlikely to return home with sums to show you. The general rule of thumb for you here is talk, talk, talk. Talk about the maths they did at school by all means, but it will probably be more productive to talk about the everyday maths that surrounds us. You know the kind of thing:

- 'If you have 4 sweets and I give you 1 (or 2 or 3...) more, how many will you have?'
- 'Shall I cut your toast in half (or quarters)? How many pieces will you have then? What happens if I cut them all in half again, to make soldiers?'

- 'Can you pretend to be launching a rocket and count backwards from 10?'
- 'Can you share out these crackers between you and your brother?'
- 'What are the different numbers on these dice? What do they all add up to?'
- 'Can you count the stairs as you go up? Can you jump up them 2 at a time?'
- 'The crisps cost 56p and the biscuits £1.25. How much money do you need?'
- 'I have £5. Is that enough to buy all 4 of you an ice cream?'
- 'It takes twice as long to get to Daniel's house as it does to get to Grandma's. It's 2 hours (or 1½ hours, or 1 hour 45 mins...) to Grandma's, so how long will it take to get to Daniel's?'

As with everything in maths, there are always lots of different ways to get to the right answer. If children start out feeling that way about numbers, they can be free to play and have a go – and not be scared of getting things wrong. By doing this, children of all abilities will work things out for themselves, and that's how we all learn and understand best.

WHEN IS THE RIGHT TIME TO START?

Being numerate is every bit as important as being literate. We all know we should read to our children and listen to the older ones as they learn, but few of us feel we know how to introduce numeracy. Fortunately, reading to very young pre-school children is an excellent start. There is plenty of research to show that children exposed to a lot of books from an early age develop higher levels of numeracy later on, no doubt because many books for pre-schoolers involve counting, recognising patterns, number rhymes and so on.

Having fun with numbers and patterns is a good start – even with very young children. Toddlers often love chanting or singing, and chanting their numbers up to 10 will always receive a round of applause! Pre-schoolers often like looking at door numbers when out walking, and spotting whether they go up in ones or not. Talk about odd numbers, even numbers, big numbers, small numbers, negative numbers, bits of numbers. It doesn't matter what. It's just conversation for now but who knows what they are absorbing and storing away for later.

MATHS MATTERS

There is plenty of research to show that adults who do not secure a basic level of numeracy will feel the negative effects on their lifestyle, their income and even their health. Those who do not succeed at a higher level will find many rewarding career paths blocked.

As you've picked up this book, you probably don't need convincing of this. We know success in maths is vital for our children, and now we need to build those feelings of success. These stem from confidence, enjoyment and a sense of purpose. We need to make sure our children are experiencing these feelings both in and out of the classroom.

I really hope this book helps you and that you enjoy it! Good luck with it all.

I.
THE NUMBER TOOL KIT

This first chapter looks at some fundamental principles of maths and introduces the tools and techniques that are used throughout the rest of this book. The other chapters allow you to dip in and out looking for the relevant subject areas, but this one includes some simple but vital principles that are important whichever aspect of maths you'd like to help your children with.

> Primary-age pupils are more likely to refer to 'numeracy' rather than to 'maths'. They may not even have heard of the word 'maths' – this doesn't matter.

Most maths lessons or numeracy hours usually start with 'mental starters'. You may also hear talk of 'mental warm-ups', 'brain-challengers', 'brain-sharpeners' and so on. This is all about getting their brains geared up and ready to tackle some numeracy.

Mental starters are based on maths with which they are familiar. They act as revision, reiteration and consolidation, as well as sending a clear signal to the brain that it should be ready to learn – something new is on the way!

A typical mental starter for a Year 3 class is:

How many 3-digit numbers can you make from the digits 2, 3 and 9 if you use each digit only once?
Answer: 6:   239, 293, 329, 392, 923 and 932.

This may be extended to ask a 'challenge question':

🎲 What is the biggest number you can make from these?
Answer: 932

This can be repeated with a different set of digits, such as *4, 6* and *8*. Children will be encouraged to use a logical approach in finding their answers. Using logic is not only an efficient method but it also ensures all answers are 'found'. The logic for this example would be along the lines of:

- Find all the numbers starting with 2, then all the numbers starting with 3, followed by all the numbers starting with 9.

It is very common for pupils to use mini whiteboards and whiteboard pens to record their mental starter answer. Children then hold these up to show the teacher before wiping them off, ready for their next answer. So if your children come home and ask for a small whiteboard on which to do their sums you now know why!

UNDERSTANDING THE BASICS
(RECEPTION AND YEARS 1 AND 2, AGES 5–7)

For children to achieve success at a starter such as the one above they must be confident about the difference between a 'digit' and a 'number'. There are 10 'digits' in our number system, namely: 0, 1, 2, 3, 4, 5, 6, 7, 8 and 9.

'Numbers' are made up by combining these digits. How digits are combined and the value of *units*, *tens* and *hundreds* and so on within a number is referred to as 'place value'.

Place Value

Place value is one of the most important concepts for children to understand. But what exactly is it? Put simply, place value is an expression to describe how we can use the 10 digits (0, 1, 2, 3, 4, 5, 6, 7, 8 and 9) in different ways to 'make' different numbers.

For example, the digit '6' on its own simply means the number 6. But if we put the digit '6' in a different 'place'... 60, the '6' now holds a different value: it no longer means 6 but 6 tens (or 60). And if we write 600, the '6' now represents 6 hundreds.

Young children are introduced to place value by learning about hundreds, tens and units (or H T U – see page 22).

In the very early years, children will start learning about number and place value by reading and writing whole numbers. They will gradually build this up so they can read and write numbers up to 100 or more, by the end of Year 2. The aim will be for them to know numbers in words as well as in figures, for example:

Figures	Words
37	thirty seven
38	thirty eight
39	thirty nine
40	forty
41	forty one
42	forty two

Writing the figures correctly does not come easily to all children. And it can take a few years before they can consistently write them the right way up and the right way round. Children need to practise writing their numbers just as much as they need to practise writing their alphabet letters.

The same methods for learning to write letters can be used for learning to write numbers, such as using tracing paper, joining up the dotted outline of the number, copying underneath... and lots and lots of practice.

Reading numbers correctly – the right way round – can also take time. When you are aged 5, distinguishing between 21 and 12 takes some thinking about.

Once children have learnt to read and write numbers, it can be all too easy to assume that they *understand* numbers. This is rarely so. Truly understanding numbers is a complex process, and there is a lot to know.

Setting out numbers in order of size is a good way to start children thinking about the meaning or 'value' of numbers. For example, if you gave your children the numbers: 4, 7, 1, 9, 2, 6, could they put them in order from smallest to biggest? This will give you an idea of whether your children are beginning to understand that numbers have value.

Using cards with numbers on them that the children can manually rearrange is often a good idea, too. Another good way to ascertain whether your children are actually beginning to *understand* numbers is to ask questions like:

- 'What is 1 more than 12?'
- 'What is 1 less than 17?'
- 'What is 10 more than 20?'
- 'What is 10 less than 25?'

Often parents will proudly say their pre-school children know all their numbers up to 20 (or 50, or 100, or whatever). But what the children may actually be doing is reciting 'sounds', parrot fashion, with no comprehension that 12 is 1 more than 11 or that 16 is 1 less than 17.

It's a bit like those cute pre-schoolers who apparently know the entire alphabet, having heard the 'ABC' song a zillion times. They can chant along happily to the music, but this doesn't imply they actually *know* their alphabet in any meaningful way. And while it is of course important for children to learn the words or sounds, we should see this for what it is – a starting point.

The Hundred Square

Another useful approach is the **Hundred Square (also known as a Hundred Number Square)**, which can really help your children understand the order of numbers. A **Hundred Square** is an arrangement of the first 100 numbers in order in a square, that is, 10 rows each with 10 numbers, like this:

1	2	3	4	5	6	7	8	9	10
11	12	13	14	15	16	17	18	19	20
21	22	23	24	25	26	27	28	29	30
31	32	33	34	35	36	37	38	39	40
41	42	43	44	45	46	47	48	49	50
51	52	53	54	55	56	57	58	59	60
61	62	63	64	65	66	67	68	69	70
71	72	73	74	75	76	77	78	79	80
81	82	83	84	85	86	87	88	89	90
91	92	93	94	95	96	97	98	99	100

Your children will undoubtedly come across these in class – there will probably be a big colourful poster of one on the wall. Children will also often be given their own individual copy to stick in their book or tray.

Number squares are very useful props for helping with counting on and for spotting patterns. They are also useful for adding, subtracting, multiplying and even dividing – more of which in later chapters.

Sticking a colourful Hundred Square up at home can really help your children. Put it somewhere they can reach, touch and point at it. Seeing which number comes next, or which number is 1 more, or 1 less, or 10 more, or 10 less – all these reminders are going to be a big help. And they'll need to keep checking and checking: seeing something once doesn't mean they know it.

Most of the big poster-style Hundred Squares also have the numbers written in words. This is really useful too.

The Number Line

Another useful way of showing the order of numbers is a **Number Line**. A Number Line looks like this:

0	1	2	3	4	5	6	7	8	9	10	11

It can be extended as far as you like in either direction. (A Number Line from 0 to 30 is typically used for pupils in Year 2.)

The Number Line is a simple but very effective tool. It is simply a list of numbers along a line. The line can be horizontal with the numbers underneath (similar to the markings on a ruler), as shown above, or vertical with the numbers to the side (like a thermometer). Often the Number Line is introduced through a simple exercise. For example, a group of children are each given a number between 1 and 10 to hold, and have to arrange themselves in order.

Initially, the Number Line might just be used to show the order of numbers and for counting on. But the Number Line has many, many uses – as you will see throughout this book.

Your children might now begin to appreciate and understand **'ordinal' numbers**. They don't need to know the word, just what it means. **Ordinal** simply gives order or position to a number. The ordinal numbers are:

- First, second, third, fourth, fifth, sixth, seventh, eighth, ninth, tenth, eleventh…

Your children will also need to know the corresponding abbreviations:

- 1st, 2nd, 3rd, 4th, 5th, 6th, 7th, 8th, 9th, 10th, 11th…

This may all seem incredibly straightforward to an adult – but remember, to a young child this is all brand new. Other words we associate with ordinal numbers are:

- Last
- Last but one
- The date is…

Questions your children might be asked in class will be along these lines:

- 'Who is the first person in the queue today?'
- 'Who was the last person ready for PE?'
- 'What is the fourth letter of the alphabet?'
- 'What date is your birthday?'
- 'If you came third in the egg-and-spoon race, how many people finished before you?'
- 'There are 9 of you in this queue. Jack, you are third in the queue. How many children are in front of you and how many behind you?'

By the end of Year 2, children will probably be using ordinal numbers up to 100.

Other numbers have names too, such as 'cardinal' or 'counting' numbers, 'nominal' numbers, 'whole' numbers, 'negative' numbers, 'integers' and so on, as explained in the box (below).

NAMING NUMBERS

Counting or 'cardinal' numbers are the numbers we use to count with! They indicate quantity and tell us 'how many'. For example: 3 raisins, 10 presents, 5 friends, 4 lunar months, 15 head of cattle, 28 days and so on. So the **counting** or **cardinal** numbers start: 1, 2, 3, 4, 5, 6...

If 0 ('zero' or 'nought') is included at the start of this list, then we have a new name for this set of numbers: **'whole numbers'**. Whole numbers start: 0, 1, 2, 3, 4, 5, 6...

If **negative numbers** are added to this list of whole numbers, then the list of numbers are known as **'integers'** (see also page 31). **Integers** are: −6, −5, −4, −3, −2, −1, 0, 1, 2, 3, 4, 5, 6...

Ordinal numbers do not show quantity or amount. Instead they indicate rank or position. For example: first; second; third; fourth...

Nominal numbers name something. We use nominal numbers for: telephone numbers, the numbers on a football shirt, the Number 4 bus, flight numbers and so on. Nominal numbers don't indicate quantity or rank order; they simply identify something.

Of course, your children do not need to know all this terminology – we just need to help them understand that we use numbers in different ways.

Your children will start to learn about **tens and units**, then later **hundreds, tens and units**. (Hundreds, Tens and Units are often referred to as **H T U**.)

For example:

16	is 1 ten and 6 units
27	is 2 tens and 7 units
73	is 7 tens and 3 units
40	is 4 tens and no units

Units are sometimes referred to as '**ones**'. They are exactly the same thing.

Children will often be introduced to these terms with the help of an abacus or other physical props.

16 = 1 **ten** and 6 **units**, and would look like this on an abacus:

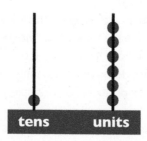

Or with interlocking cubes or Cuisenaire rods – remember them? Little sticks of different sizes with markings divided into 1cm squares – and would look like this:

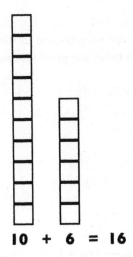

10 + 6 = 16

Children may also be given a table like the one below and asked to put the digits in the correct columns. For example they might be asked to write 15, 8, 36, 40 and 55 in a table like this:

T	U
1	5
	8
3	6
4	0
5	5

This will be extended to include hundreds. For example, they might be asked to write 165, 38, 9, 360, 400 and 505 in a table like this:

H	T	U
1	6	5
	3	8
		9
3	6	0
4	0	0
5	0	5

The **HTU** table can be extended to the left indefinitely, each column being ten times the previous one:

H of Th	T of Th	Th	H	T	U
Hundreds of thousands	Tens of thousands	Thousands	Hundreds	Tens	Units
100 000	10 000	1000	100	10	1
←	×10	×10 ×10	×10	×10	←

Numbers are also referred to as 1-digit, 2-digit or 3-digit numbers. For example:

1-digit: 2 or 8 or 6 or 3 or...   (just **units**)
2-digit: 13 or 37 or 99 or 83 or...   (**tens** and **units**)
3-digit: 238 or 472 or 791 or...   (**hundreds**, **tens** and **units**)

0 ('zero' or 'nought') can act as a **place holder**. All that means is that in the number '40', for example, we put a 0 in the units column, which then allows the 4 to be 'held' in the right position to show that it represents 4 **tens**.

T	U
4	0

When the concept of '0' (or zero) was first introduced, some 3000 years ago, it was only regarded as a place holder or separator. Now 0 as a number and 0 as a place holder are interchangeable. In the number 306, for example, the 0 in the middle not only allows the 3 to be in the right position or place, but also tells us that we have 0 – or no – **tens**.

As your children begin to appreciate the value of a number, they will start to learn how to split or **partition** numbers into **tens** and **units**. This helps to reinforce the notion of which 'bit' is the tens and which 'bit' is the **units**, thereby helping with reading and understanding numbers.

For example:

16	=	10	+	6
12	=	10	+	2
21	=	20	+	1
83	=	80	+	3

Children will also be introduced to language such as **more** or **less** and **larger** or **smaller**, to compare two numbers. Questions like the ones below can help children appreciate these words in a mathematical sense:

- 'Which is less: 12p or 21p?'
- 'Which is more: 14 or 18?'
- 'Who has the larger number of beads – you or Jemima?'
- 'Which is smaller: 25 or 52?'

Following on from this, children will then be asked to say a number that lies **between** two given numbers.

- 'Can you think of a number between 28 and 32?'
- 'Are there any other numbers between 28 and 32?'
- 'What number is halfway between 10 and 20?'
- 'What number is halfway between 19 and 23?'

A good game to play with your children is '*Guess my number*'. Two people play this game and you can take it in turns to start. The first person thinks of a number and starts by saying '*My number is between…*' and then must give only '*Yes*' or '*No*' answers to the other person's questions.

For example Ruby and Eden play a game:

Ruby:	'My number is between 20 and 100'
Eden:	'Is it less than 60?'
Ruby:	'Yes'

Eden:	'Is it more than 40?'
Ruby:	'No'
Eden:	'Is it more than 30?'
Ruby:	'Yes'

And so it goes on.

Obviously this game can be adapted for any age child. Very young children can play with an adult using smaller numbers with a smaller gap between them. Larger numbers are good for those who have mastered the idea. You can go on to challenge your children to guess the mystery number with the fewest questions.

As well as using the '=' sign to represent equality (that two amounts are the same) some children in Year 2 will be introduced to the **more than** ('>') and **less than** ('<') signs. For example:

- 8 > 3 means '8 is more than (or bigger or greater than) 3'
- 2 < 7 means '2 is less than (or smaller or fewer than) 7'

Estimating and rounding

Estimating is another fundamentally important skill. As well as being an important technique in maths, being able to estimate is a really useful tool in everyday life. Things we might need to estimate include:

- amounts of money
- periods of time
- distances
- heights/lengths/weights – of people, animals or objects

When we go to the corner shop for milk and a newspaper, we need to know, roughly, if we have enough money. In the January sales, it is good to know, roughly, how much money we have 'saved'. On the long drive to grandma's house it is helpful to know, roughly, *'how much longer?'* it will take to get there, when you've been asked for about the hundredth time by a bored 6 year old.

Being able to estimate well and having a 'feel for numbers' go hand in hand. It's about having an awareness of the world around us and being able to see the 'big picture'.

Estimating in its very early form is about guessing (for example, guessing how many objects you can see) and then checking that guess by counting.

With practice the 'guessing' becomes more accurate – especially as your children build up strategies to help them. They may start to group objects mentally in 'lots of' 5 or 10 when estimating 'how many?' If they 'know' how long a favourite television programme lasts, they can use this to develop a 'feel' for how long it takes to get to grandma's house if you explain you could watch 4 episodes during the journey.

Other words and phrases associated with estimating that your child may hear include:

- roughly, nearly, about the same as, enough, not enough, close to, too many, too few, just over, just under, approximately, nearest and around.

Giving a sensible estimate of up to 50 or so objects would be a good goal for Year 2 pupils. Younger children would start by estimating fewer objects.

Your children might also be asked to estimate the position of a point on a Number Line. For example:

Estimate the number marked by the arrow on the Number Line below. How did you decide?

$$\downarrow$$

0	10

The strategy your children will probably employ would be to look for the halfway point and decide whether the arrow is *'close to'*, *'about the same as'*, *'just under'*, or *'just over'* this point. In the example above, a good estimate would be 4.

Rounding is another skill that children need to learn. Like **estimating** it can help to make numbers easier to manage. In Year 2, children will begin to learn how to round numbers (less than 100) to the nearest 10.

For example:

🎲 Round the following numbers to the nearest 10:
23, 48, 35.

First, a child would need to know how to count up in tens:

10, 20, 30, 40, 50, 60, 70...

Now let's look at each number in turn.

🎲 23 is more than 20 and less than 30 – but which one is it closer to?
Answer: 20

🎲 48 is more than 40 but less than 50, but which one is it closer to?
Answer: 50

🎲 35 is more than 30 but less than 40, but which one is it closer to? Well this one is a bit trickier, because 35 is exactly halfway between 30 and 40. So which one do we round it to? Well, we always **round up** if the number is exactly in the middle. So...
Answer: 40

MOVING FORWARD
(YEARS 3 AND 4, AGES 7–9)

In this chapter, as in all chapters, as we move from one section to the next, it is worth bearing in mind the following. Just because a child has been taught various ideas and concepts it does not follow that the child *understands* these ideas and concepts. Much reiteration and revision will continue to take place alongside the introduction of new material. In this section we see the tools and techniques that children use to build on from the basics set out in the previous **'Understanding the Basics'** section.

A **counting stick** is a piece of equipment that some teachers like to use. You are just as likely to see it in a Year 4 class as in a Year 7 class. It is quite literally a stick, about one metre in length, divided into 10 sections with alternate sections shaded:

The stick can be used in a multitude of ways. For example, one end may represent 0, and the other end 10. Children may then be asked to point to 7.

One end may represent 0 and the other end 100. Children may then be asked to '*identify 80*'.

The far end may even represent 1000. In that case, '*where would 600 be?*'

The starting number may be changed to 13, for example, and the other end to 23. In that case, '*where would 19 be?*'

Or the ends may represent 482 and 582 respectively, or 7865 and 8865.

Decimal numbers can also be explained using a counting stick. One end could be 0 and the other end 1.

🔲 What does each interval represent now?
 Answer: 0.1

🔲 Where is halfway on the stick? What decimal number is this?
 Answer: 0.5

Year 3 children will be expected to read and write whole numbers to at least 1000 (one thousand) in figures and in words. By the end of Year 4 some children will be reading and writing numbers well beyond 10 000 (ten thousand).

This is great progress, but being able to read and write numbers is not enough. They will also need to appreciate **place value** and so know what each digit represents (see also page 16).

Just as we saw above, a number can be split or **partitioned** into its component parts. Now children will start to partition into **thousands, hundreds, tens** and **units** (**Th, H, T** and **U**). This helps with reading and understanding numbers.

For example:

14	=					10	+	4
56	=					50	+	6
99	=					90	+	9
125	=			100	+	20	+	5
586	=			500	+	80	+	6
803	=			800	+	0	+	3
900	=			900	+	0	+	0
999	=			900	+	90	+	9
1234	=	1000	+	200	+	30	+	4
3875	=	3000	+	800	+	70	+	5

A good exercise when practising **place value** – and one that most pupils find fun – is to hand your children 3 single-digit number cards, such as: **3, 8 and 2** and ask them to make the biggest number they can by rearranging the cards. *'Can they say that number? Now find the smallest number.'*

- In this case, the biggest number would be: 832
  (eight hundred and thirty-two)
- And the smallest number would be: 238
  (two hundred and thirty-eight)

This really helps children appreciate that where you *place* a digit has a big impact on the size or *value* of a number.

A typical exercise for a Year 4 (or Year 5 or Year 6) class is:

🔷 There are 10 digits in total: 0, 1, 2, 3, 4, 5, 6, 7, 8, 9. Without using any digit more than once in each sum: What is the biggest 4-digit sum you can make? What is the smallest 4-digit sum you can make? Investigate.

Possible answers (there are others, although the sum total will be the same):

Biggest:	Smallest:	Or is the smallest?:
9 7 5 3	1 0 4 6	0 2 4 6
+ 8 6 4 2	+ 2 3 5 7	+ 1 3 5 7
—————	————	—————
1 8 3 9 5	3 4 0 3	1 6 0 3

> Integer is just another name for a whole number. (Children *do* need to be familiar with this word.)

Multiplying an integer by 10 or 100

Multiplying an integer by 10 or 100 is a skill children will be taught. But how do we calculate this?

In your head, how would *you* do this multiplication:

24 × 10?

Or this one:

38 × 100?

If you are not sure, don't worry – it is all explained below.

If you did these sums in your head, did you *'just add a nought when you multiplied by 10, and 2 noughts when you multiplied by 100'* giving the answers 240 and 3800 respectively? I did, and generally would do this – but are we 'allowed' to tell our children to do likewise?

Technically speaking, the answer is 'no'. And it is not what they will be taught in class, at least not at first.

Why not? The reason is that this method falls down badly when multiplying by anything *other* than an integer (whole number).

Look at these:

2.4	×	10	=	24
0.38	×	100	=	38
2¼	×	10	=	22½

Simply adding a 0 or two would not work here!

So what are children taught? They may very well be shown a grid like the one below and asked to describe what they 'see' or notice. Each line, going down the grid, shows a multiplication by 10.

1	2	3	4	5	6	7	8	9
10	20	30	40	50	60	70	80	90
100	200	300	400	500	600	700	800	900
1000	2000	3000	4000	5000	6000	7000	8000	9000

The idea would be for children to see and understand that:

- When an integer is multiplied by 10, the *digits* move **one** place to the **left**.
- Multiplying by 100 is the same as multiplying by 10 and then by 10 again. So the *digits* move **two** places to the **left**.

This can also be shown in a **HTU** table:

For example: 28 × 10:

H	T	U
	2	8
2	8	0

The 0 is 'put' in the units column to act as the place holder.

For example: 28 × 100:

0 is 'put' in the tens and also in the units column as place holders.

This method will work just the same if we are multiplying decimal numbers by 10 or 100, as long as we understand a couple of key rules as outlined in the 'To the Top' section below (see also pages 167 and 189).

Dividing an integer by 10 or 100

It's important to remember that division is the opposite of – or will 'undo' – multiplication. Therefore dividing by an integer is virtually identical to the technique for multiplication shown above – only the reverse.

So, looking again at the grid we saw earlier:

1	2	3	4	5	6	7	8	9
10	20	30	40	50	60	70	80	90
100	200	300	400	500	600	700	800	900
1000	2000	3000	4000	5000	6000	7000	8000	9000

We now start at the bottom of the grid instead and go up. Each line involves a division by 10. The idea would be for children to see and understand that:

- When a whole number is divided by 10, the *digits* move **one** place to the **right**.
- Dividing by 100 is the same as dividing by 10 and then by 10 again. So *digits* move **two** places to the **right**.

This can also be shown in a **HTU** table:
For example: 2800 ÷ 10:

For example: 2800 ÷ 100:

Th	H	T	U
2	8	0	0
		2	8

Decimal notation for tenths and hundredths will be introduced to *some* children as early as Year 3 or Year 4.

Just as the **HTU** table can be extended to the left indefinitely, so it can also be extended to the right indefinitely. A decimal point separates the units from these columns and then each column to the right is 10 times smaller, or a tenth, of the previous one. This is equivalent to dividing by 10.

You'll notice that **tenths, hundredths** and **thousandths** are abbreviated to **t, h and th**.

H	T	U	.	t	h	th
Hundreds	Tens	Units	Decimal point	tenths	hundredths	thousandths
100	10	1	.	$\frac{1}{10}$	$\frac{1}{100}$	$\frac{1}{1000}$
÷ 10	÷10		÷ 10	÷ 10	÷ 10	→

The notion of tenths and hundredths may be introduced in relation to money and measurement at first.

For example, children might be asked to write these amounts in a table: £2.50, £15.25, £4.05, £124, £230.50

H	T	U	.	t	h
		2	.	5	0
	1	5	.	2	5
		4	.	0	5
1	2	4	.	0	0
2	3	0	.	5	0

Or they might be asked questions such as:

- 'Which is more: £2.50 or £2.05?'
- 'Why?'
- 'Which is longer: 1.50m or 1.05m?'
- 'Why?'

Using 'money' or 'measurements' as a way of introducing decimal numbers is sensible, as it relates the maths lesson to the real world. Children are then more likely to understand what they are being taught because they are familiar with the terminology. However, there are limitations to this approach (see page 46).

If children only see decimal numbers as amounts of money or as measurements, they can become resistant to the idea of decimal numbers looking like anything else, such as:

2.5643 or 74.3333333333 or 0.00005

Numbers smaller or less than 0 are called **negative numbers**, while those bigger than 0 are called **positive numbers**.

Negative numbers

Negative numbers will probably be introduced some time around Year 4. Negative numbers are shown with a minus sign ('−') as follows:

−1, −2, −3, −4 and so on.

We say these as '*negative one*', '*negative two*', '*negative three*', '*negative four*' and so on, or as '*minus one*', '*minus two*', '*minus three*', '*minus four*' and so forth.

'Negative one' is 1 less than 0, 'negative two' is 2 less than 0, and so on. What this means (and this is the bit children sometimes take a while to grasp) is that 'negative six' is **smaller** than 'negative five'.

A Number Line can be used to help make this clearer. As we have seen above, the Number Line can be extended as far as we like. Not only can it go on and on and on as numbers get larger, but it can be extended in the other direction too.

| -6 | -5 | -4 | -3 | -2 | -1 | 0 | 1 | 2 | 3 | 4 |

Here we can see that –6 is to the left of –5, so it *must* be smaller.

Counting backwards is a good way for your children to become familiar with negative numbers. Count with them, just as you did when your children were small: 10, 9, 8, 7, 6, 5, 4, 3, 2, 1… but this time don't stop… 0, –1, –2, –3, –4, –5, –6 and so on.

In fact, I would like to suggest that this counting backwards exercise is done sooner rather than later. Don't be afraid to introduce it when your children are tiny. The minds of small children are open and receptive. They are learning so much every day that they have yet to become resistant to new or unfamiliar material. Basic familiarity is all you should hope to achieve; a sense of knowing that, for example, –4 comes after (and is therefore smaller than) –3.

Another helpful device here can be to think about the weather and the temperature gauge. If we hear on the weather forecast that it's going to be 'minus 7 degrees tonight' we know it's going to be pretty cold, a lot colder than 'minus 1', for example. This also ties in with the type of questions your children are likely to be asked in class:

- 'Which temperature is lower: −5 or −2?'
- 'What is the coldest temperature: −12 or −15?'
- 'Can you put the following temperatures in order, lowest to highest: 3   −2   5   −8   −6'
- 'What integers lie between −4 and 2?'
- 'Fill in the missing numbers on the Number Line below.'

| -6 | -5 | ... | -3 | ... | -1 | ... | 1 | 2 | ... | 4 |

Negative numbers are sometimes related to debt and owing money.

For example: +4 might be related to us *owning £4* while –4 might be related to us *owing £4* (that is a debt of £4).

Children will also learn to add, subtract, multiply and divide with negative numbers. This will be explained in the next four chapters.

By this stage children may already have been introduced to the **inequality symbols: more than ('>')** and **less than ('<')**. Now they need to use them correctly in their written work.

For example:

🎲 If $2560 < \square < 2580$, what number could $\square$ be?

🎲 Fill in the missing symbol to make the statement correct: 3950 ... 3874.

🎲 Is $-3 > -5$ true or false?

🎲 Is $-1 < 1$?

The answers are:
- Anything from 2561, 2562, 2563, 2564... up to 2579.
- $3950 > 3874$
- True
- Yes

Estimating and rounding

Building on from the basics, children will continue to practise **estimating** but with bigger quantities and larger measurements.

Examples of questions your children might encounter are:

- 'This jar can hold 100 sweets when it is full. Some have been eaten. Can you estimate how many might be left?'
- 'Can you estimate how many words there are on this page?'
- 'Estimate the number marked by the arrow on the Number Line below. How did you decide?'

$$\downarrow$$

---

0   100

- 'Estimate the height of the classroom.'

- 'Estimate the length of 1 minute. I will tell you when to start. Close your eyes and put your hands up when you think a minute has passed.'

This last one is quite entertaining to do. You might like to try it at home with your own children. Most children – even those of secondary school age – are notoriously bad at estimating time. They think it goes much quicker than it does!

Your children will need to discover strategies to help them estimate. Some strategies to help with the questions above include:

- Working out how many sweets are on one 'layer' and then roughly calculating the probable number of 'layers'.
- A quick count of how many words there are on a typical-looking line of the page and then roughly calculating how many lines there are on the page.
- Looking for the halfway point on the Number Line and considering how close to this point the arrow lies.
- Knowing roughly what 1 metre is (a child's long stride is often quite close to 1 metre) and using this to estimate how many metres could go up the side of the classroom.
- Teaching your children how long 1 second is. It usually takes 1 second to say, out loud, the words: 'One elephant.' To estimate a minute they need to count up to 60 elephants: '1 elephant, 2 elephants, 3 elephants, 4 elephants... 59 elephants, 60 elephants!' It surprises most children how long a minute actually is.

Children will also be encouraged – more and more – to estimate answers to sums before they attempt to solve them 'properly'. In this way children will have a rough idea of what they expect the answer to be before they begin. Once they have completed their calculation they can then compare their answer to their 'estimate' and ask the very important question: '*Is my answer sensible?*'

Children of all ages are encouraged to check their work in this way. This is echoed throughout this book.

Rounding is another very important skill and is used when we don't need to know, or indeed don't *want* to know, exact numbers. For example:

- Ben has £40 birthday money. He wants to buy 3 things with this money. They are a DVD (for £16.99), a glow-in-the-dark frisbee (for £11.25) and a model spaceship (for £8.50). Ben wants to know whether he can get all 3 now.

By rounding the amounts to the nearest pound, Ben quickly works out that he *can* get everything he wants: £17 + £11 + £9 = £37.

Building on from rounding 2-digit numbers to the nearest 10 (see page 28), we now extend this technique to rounding bigger numbers to the nearest 10 and then rounding to the nearest 100.

As we saw earlier, children need to practise until they can count up in tens confidently: 10, 20, 30, 40, 50, 60, 70, 80, 90, 100, 110, 120, 130...190, 200, 210... 290, 300, 310... and so on.

Likewise in hundreds: 100, 200, 300, 400, 500, 600, 700...

Then rounding becomes easy.

Let's look at how to round the following numbers, first by rounding to the nearest 10 and then by rounding to the nearest 100:

533
647
875
250

Rounding to the nearest 10.

---

533 is more than 530 but less than 540, but which one is it closer to?
Answer: 530

---

647 is more than 640 but less than 650, but which one is it closer to?
Answer: 650

● 875 is more than 870 but less than 880, but which one is it closer to?

This one is a bit trickier, because 875 is exactly halfway between 870 and 880. Remember, we always **round up** if the number is exactly in the middle. So which one do we round it to?
Answer: 880

● 250 is already a multiple of 10, so no rounding is necessary.
Answer: 250

Rounding to the nearest 100.

● 533 is more than 500 but less than 600, but which one is it closer to?
Answer: 500

● 647 is more than 600 but less than 700, but which one is it closer to?
Answer: 600

● 875 is more than 800 but less than 900, but which one is it closer to?
Answer: 900

● 250 is more than 200 but less than 300, but which one is it closer to?

This one looks trickier, because 250 is exactly halfway between 200 and 300. So which one do we round it to? Again, we **round up** if the number is exactly in the middle.
Answer: 300

Rounding in this way is often used to find approximate solutions to addition and subtraction sums.

For example:

- 'Which of these is the best approximation for 207 + 586?
  200 + 500 • 300 + 600 • 200 + 600 • 200 + 500'

- 'Which of these is the best approximation for 895 − 602?
  900 − 700 • 800 − 600 • 800 − 700 • 900 − 600'

Similarly, rounding is used to find approximate solutions to multiplication and division problems:

For example:

- 'Which of these is the best approximation for 19 × 32?
  20 × 30 • 10 × 30 • 10 × 40 • 20 × 40'
- 'Which of these is the best approximation for 158 ÷ 41?
  150 ÷ 50 • 150 ÷ 40 • 160 ÷ 40 • 200 ÷ 40'

So rounding can be used to help give quick and efficient 'answers'.

The answers to the above are, respectively:

200 + 600	giving an approximate answer of 800
900 − 600	giving an approximate answer of 300
20 × 30	giving an approximate answer of 600
160 ÷ 40	giving an approximate answer of 4

TO THE TOP
(YEARS 5 AND 6 PLUS, AGES 9–12 PLUS)

Children in Years 5, 6, 7 and beyond will now read and write numbers in both figures and words, knowing what each digit represents in whole numbers and in decimal numbers with up to 3 decimal places. They will also continue to **multiply by 10 or 100.**

Following on from the introduction given in the '**Moving Forward**' section, we can now see how we use the same method to multiply decimal numbers by 10 or 100. Once again the numbers can be first shown in a table and then the same rules as we saw earlier are applied:

- When a number is multiplied by 10 the *digits* move **one** place to the **left.**
- Multiplying by 100 is the same as multiplying by 10 and then by 10 again. So *digits* move **two** places to the **left.**

For example: 7.4 × 10

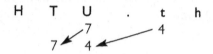

Each digit has moved **one** place to the **left** giving the answer:

7.4 × 10 = 74

For example: 7.4 × 100

H   T   U   .   t   h
7   ↙7   ─── 4
7 ↙ 4 ↙ 0

Each digit has now moved **two** places to the **left**, and a 0 is 'put' in the **units** column to act as a place holder, giving the answer:

7.4 × 100 = 740 (see also page 189).

Children will continue to multiply **integers** (whole numbers) by 10 and 100 but will now extend this to **multiplying by 1000.**

- Multiplying by 1000 is the same as multiplying by 10, then by 10 again and then by 10 again. So *digits* move three places to the left.

For example: 34 × 1000:

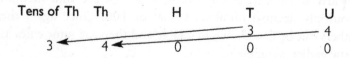

Each digit has now moved **three** places to the **left** and a 0 is 'put' in the hundreds, tens and units columns to act as place holders, giving the answer:

34 × 1000 = 34000

Children will also continue to **divide by 10 or 100**. Again, we can use the same method to divide decimal numbers by 10 or 100. Here too the numbers can be first shown in a table and then the same rules are applied:

- When a number is divided by 10, the *digits* move **one** place to the **right**.
- Dividing by 100 is the same as dividing by 10 then by 10 again. So *digits* move **two** places to the **right**.

For example: **25 ÷ 10**

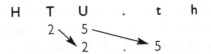

Each digit has moved **one** place to the **right** giving the answer:

25 ÷ 10 = 2.5

For example: **25 ÷ 100**

<pre>
H   T   U   .   t   h
      2   5
            0   .   2   5
</pre>

Each digit has now moved **two** places to the **right**, and a 0 is 'put' in the **units** column, giving the answer:

25 ÷ 100 = 0.25 (see also page 224)

Children will continue to divide integers (whole numbers) by 10 and 100 but will now extend this to **dividing by 1000**.

- Dividing by 1000 is the same as dividing by 10, then by 10 again, and then by 10 again. So the *digits* move three places to the right.

For example: 26000 ÷ 1000:

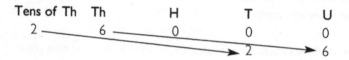

Tens of Th	Th	H	T	U
2	6	0	0	0
			2	6

Each digit has now moved **three** places to the **right**, giving the answer:

26000 ÷ 1000 = 26

Decimal notation for tenths and hundredths was introduced in the earlier section. Now children might be expected to use **decimal notation for tenths, hundredths and thousandths**:

H	T	U	.	t	h	th
Hundreds	Tens	Units	Decimal point	tenths	hundredths	thousandths
100	10	1	.	$\frac{1}{10}$	$\frac{1}{100}$	$\frac{1}{1000}$

Just as we saw above, a number can be split or **partitioned** into its component parts. Now children will start to partition decimal numbers.

For example:

51.4	=	50	+	1	+	0.4			
2.67	=			2	+	0.6	+	0.07	
34.902	=	30	+	4	+	0.9	+	0.00	+	0.002

This can prove quite tricky for quite a lot of children initially, so is well worth practising. To test your children's understanding, you could try asking questions such as:

⬢ Can you put these numbers in order, smallest to largest:
   4.2 • 14.2 • 4.25 • 4.09 • 4.225 • 4.100

⬢ List these numbers from smallest to largest:
   5.65 • 5.600 • 6.5 • 5.675 • 5.765 • 5.7

The correct answers are:
   4.09 • 4.100 • 4.2 • 4.225 • 4.25 • 14.2
   5.600 • 5.65 • 5.675 • 5.7 • 5.765 • 6.5

Mistakes will be made if children have not fully grasped place value with decimal numbers. It's not always easy to grasp – children may take a long time and need a lot of practice and be well into secondary school before they are fully competent.

A very common mistake is for children to think that, for example, 5.7 is smaller than 5.600. This may be because 5.7 is a shorter number to write, and when we are not absolutely sure of ourselves, our brains will revert to what we already 'know' and try to 'fit' the problem to something with which we are familiar and comfortable.

Children can compare the two numbers, look for differences and 'see' that to the right of the decimal point we have a choice of 7 or 600. In everything children have ever done up to this point, 600 has unquestionably been bigger than 7, so it's no wonder if they conclude that 5.600 is bigger than 5.7.

This is why it's so important – and why teachers stress the point so much – for children (and adults) to *say* numbers correctly. For example, 5.600 is '*five point six zero zero*'. And not '*five point six hundred*'!

Each digit after a decimal point must be pronounced separately.

However, there are exceptions to this rule, and that is where confusion can creep in. As we saw in the last section, often children are introduced to decimal numbers (with tenths and hundredths) by way of money examples. This is a sensible approach as most children are familiar with money and it puts the maths in context in the real world.

Unfortunately, with money we generally don't pronounce our decimal numbers correctly. So for £5.65, we say *'five pounds sixty-five'* and not *'five point six five pounds'* as perhaps we should. What we are actually doing when we say *'five pounds sixty-five'* is noting that it is 5 pounds and 65 pence.

By highlighting this obvious exception as early as possible, we may be able to avoid potential confusion.

Children will be expected to count using numbers up to a **million** at this stage. A million is 'one thousand lots of one thousand' and looks like this:

1 000 000

In a table one million would look like:

M	H of Th	T of Th	Th	H	T	U
Million	Hundreds of thousands	Tens of thousands	Thousands	Hundreds	Tens	Units
1	0	0	0	0	0	0

It is the same as $1 \times 10 \times 10 \times 10 \times 10 \times 10 \times 10$.

WHAT IS A BILLION?

This is not as easy to answer as you might think. A billion used to be a million lots of a million and looked like this:

1 000 000 000 000

and this is still the case in large parts of Europe. But in the US system, a billion is a thousand lots of a million and looks like this:

1 000 000 000

Britain has now largely adopted the US system. The same incon-sistency is true of trillions. (And trillions always seem to be on the news these days.) For Americans, a trillion is a thousand (of their) billion and looks like this:

1 000 000 000 000

Which is in fact the same as the European billion. A European tril-lion is a million million million and looks like this:

1 000 000 000 000 000 000

It's all very confusing isn't it? A table might help clarify matters:

	USA and scientific community	Other countries
1 000	thousand	thousand
1 000 000	million	million
1 000 000 000	billion	1000 million (1 milliard)
1 000 000 000 000	trillion	billion
1 000 000 000 000 000	quadrillion	1000 billion
1 000 000 000 000 000 000	quintillion	trillion

In the UK the US trillion is now widely used.

The inequality symbols of **more than** ('>') and **less than** ('<') may well have been introduced already (see also page 26). Now chil-dren may also be introduced to the symbols:

- **more than** or **equal to** ('≥')
and
- **less than** or **equal to** ('≤')

For example:

List all the whole numbers for which this inequality holds true:
$$10 \leq \square \leq 15$$
So $\square$ must be a number somewhere between 10 and 15 but could also be equal to 10 or 15.
Answer: 10, 11, 12, 13, 14 or 15

Estimating and rounding

Estimating continues to be a very important and fundamental maths skill. Children will continue to use 'estimation' throughout Years 5, 6 7, 8 and beyond.

With more practice and more awareness of the world around them, children can become expert estimators. There are endless opportunities to estimate.

For example:

- 'Estimate how many biscuits there are in this packet. Do you think there will be enough for the whole class?'
- 'Estimate how many pints of milk your family needs in a week.'
- 'Estimate how many slices of bread you eat in a week.'
- 'Estimate how much money you need to...'

Rounding is now extended to rounding any number (up to 10000) to the nearest 10, the nearest 100 and also to the nearest 1000. Just as we saw earlier, to make this as easy as possible, children must feel confident about counting up accurately in tens:

10, 20, 30...

And in hundreds:

100, 200, 300, 400, 500, 600…

And in thousands:

1000, 2000, 3000, 4000, 5000, 6000, 7000, 8000…

Let's look at rounding the following numbers, first rounding to the nearest 10, then to the nearest 100, and finally to the nearest 1000:

7463 • 4647 • 2735 • 7850 • 2500

Rounding to the nearest 10.

● 7463 is more than 7460 but less than 7470, but which one is it closer to?
Answer: 7460

● 4647 is more than 4640 but less than 4650, but which one is it closer to?
Answer: 4650

● 2735 is more than 2730 but less than 2740, but which one is it closer to? Well this one is a bit trickier, because 2735 is exactly halfway between 2730 and 2740. So which one do we round it to? As we have seen previously, we round up if the number is exactly in the middle.
Answer: 2740

● 7850 is already a multiple of 10, so no rounding is necessary.
Answer: 7850

● 2500 is already a multiple of 10, so no rounding is necessary.
Answer: 2500

Rounding to the nearest 100.

7463 is more than 7400 but less than 7500, but which one is it closer to?
Answer: 7500

4647 is more than 4600 but less than 4700, but which one is it closer to?
Answer: 4600

2735 is more than 2700 but less than 2800, but which one is it closer to?
Answer: 2700

7850 is more than 7800 but less than 7900, but which one is it closer to?
As 7850 is exactly halfway between 7800 and 7900 we round up.
Answer: 7900

2500 is all ready a multiple of 100, so no rounding is necessary.
Answer: 2500

Rounding to the nearest 1000.

7463 is more than 7000 but less than 8000, but which one is it closer to?
Answer: 7000

4647 is more than 4000 but less than 5000, but which one is it closer to?
Answer: 5000

2735 is more than 2000 but less than 3000, but which one is it closer to?
Answer: 3000

⬤ 7850 is more than 7000 but less than 8000, but which one is it closer to?
Answer: 8000

⬤ 2500 is more than 2000 but less than 3000, but which one is it closer to?
As 2500 is exactly halfway between 2000 and 3000 we round up.
Answer: 3000

Rounding is now extended even further to **rounding decimal numbers**.

First, children will be taught to round decimal numbers to the nearest whole number, and then later to the nearest tenth.

This is where things can really get a bit tricky and where some children can come unstuck if they don't have a rock solid understanding of 'place value'. So if you are at all unsure of your children's (or your own) place value credentials, it might be worth going back over earlier parts of this chapter.

For example, we will round the following decimal numbers to the nearest whole number:

8.7 • 24.3 • 18.5 • 32.68 • 57.39 • 237.61

⬤ 8.7 is more than 8 but less than 9, but which one is it closer to? Well, we look at the first digit to the right of the decimal point and see there are 7 tenths. It may help your children to see this number written in a table:

H	T	U	.	t	h
		8	.	7	

Or for your children to visualise the number on a Number Line:

8 _____|_____ 9
                        8.7

Answer: 9

24.3 is more than 24 but less than 25, but which one is it closer to? Well, we look at the first digit to the right of the decimal point and see there are 3 tenths. It may help your children to see this number written in a table:

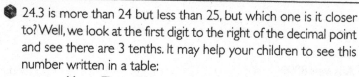

H	T	U	.	t	h
	2	4	.	3	

Or for your children to visualise the number on a Number Line:

24
24.3
25

Answer: 24

18.5 is more than 18 but less than 19, but which one is it closer to? Well we look at the first digit to the right of the decimal point and see there are 5 tenths. It may help your children to see this number written in a table:

H	T	U	.	t	h
	1	8	.	5	

Or for your children to visualise the number on a Number Line:

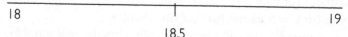

18
18.5
19

And 18.5 is exactly halfway between 18 and 19. So which one do we round it to? We always round up if the number is exactly in the middle.

Answer: 19

32.68 is more than 32 but less than 33, but which one is it closer to? Now there are 2 digits to the right of the decimal point, but we still look at the first digit and see there are 6 tenths. It may help your children to see this number written in a table:

H	T	U	.	t	h
	3	2	.	6	8

Or for your children to visualise the number on a Number Line:

32
32.68
33

Answer: 33

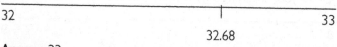

57.39 is more than 57 but less than 58, but which one is it closer to? Here, too, there are 2 digits to the right of the decimal point, but we still look at the first and see there are 3 tenths. Again, it may help your children to see this number written in a table:

H	T	U	.	t	h
5	7	.		3	9

Or for your children to visualise the number on a Number Line:

57   57.39   58

Answer: 57

237.61 is more than 237 but less than 238, but which one is it closer to? Well now there are 2 digits to the right of the decimal point but we still look at the first and see there are 6 tenths. Let's look at this number written in a table:

H	T	U	.	t	h
2	3	7	.	6	1

Or visualise the number on a Number Line:

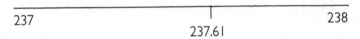

237   237.61   238

Answer: 238

So to round decimal numbers to the nearest whole number, we only look at the first digit to the right of the decimal point to see how many tenths there are. If there are 5 or more tenths we round the number up.

When children can round decimal numbers to the nearest whole number with confidence, they may be ready to **round decimal numbers to the nearest tenth**. This really challenges their understanding of place value and rounding.

For example, we will round the following decimal numbers to the nearest tenth:

9.27 • 24.84 • 8.65 • 34.50 • 34.05 • 10.123

An alternative expression for 'rounding to the nearest tenth' is **'rounding to 1 decimal place'**.

9.27 is more than 9.2 but less than 9.3, but which one is it closer to? Well we look at the second digit to the right of the decimal point and see there are 7 hundredths. It may help your children to see this number written in a table:

H	T	U	.	t	h
		9	.	2	7

Or for your children to visualise the number on a Number Line:

9.2	9.27	9.3

Answer: 9.3

24.84 is more than 24.8 but less than 24.9, but which one is it closer to? We look at the second digit to the right of the decimal point and see there are 4 hundredths. It may help your children to see this number written in a table:

H	T	U	.	t	h
	2	4	.	8	4

Or visualise the number on a Number Line:

24.8	24.84	24.9

Answer: 24.8

8.65 is more than 8.6 but less than 8.7, but which one is it closer to? Well we look at the second digit to the right of the decimal point and see there are 5 hundredths. It may help your children to see this number written in a table:

H	T	U	.	t	h
		8	.	6	5

See the number on a Number Line:

8.6	8.65	8.7

And 8.65 is exactly halfway between 8.6 and 8.7. So which one do we round it to? We always **round up** if the number is exactly in the middle.
Answer: 8.7

🎲 34.50 is already exactly the same as 34.5. It may help your children to see this number written in a table:

H	T	U	.	t	h
	3	4	.	5	0

Answer: 34.5

🎲 34.05 is more than 34.0 but less than 34.1, but which one is it closer to? Well we look at the second digit to the right of the decimal point and see there are 5 hundredths. It may help your children to see this number written in a table:

H	T	U	.	t	h
	3	4	.	0	5

Or for your children to visualise the number on a Number Line:

```
34.0 ─────────────┬───────────────── 34.1
                34.05
```

And 34.05 is exactly halfway between 34.0 and 34.1. So which one do we round it to? We always **round up** if the number is exactly in the middle.
Answer: 34.1

🎲 6.823 is bigger than 6.8 but less than 6.9, but which one is it closer to? There are 3 digits to the right of the decimal point but we still look at the second and see there are 2 hundredths. Show this number written in a table:

H	T	U	.	t	h	th
		6	.	8	2	3

Or visualise the number on a Number Line:

```
6.8 ──┬───────────────────────────── 6.9
    6.823
```

Answer: 6.8

So, to round decimal numbers to the nearest tenth of a number (or 1 decimal place), we look at the **second** digit to the right of the decimal point to see how many **hundredths** there are. If there are 5 or more hundredths we round the number up.

There is one example of rounding that does cause endless problems.

First look at these examples of rounding to the nearest tenth:

4.66 → 4.7   (to 1 decimal place)
4.76 → 4.8   (to 1 decimal place)
4.86 → 4.9   (to 1 decimal place)

Now consider rounding 4.96 to the nearest tenth. Well, as with the examples above, we look at the second digit to the right of the decimal place and see that it is 6 hundredths, so we want to round up to the next tenth. But what is the next tenth after 4.9? **Answer:** 4.96 → 5.0 (to 1 decimal place)

This type of question causes problems because as a result of rounding to the next tenth the digit in the units column needs to change. This relies on children understanding increments of 0.1 (for example, 4.9 + 0.1 (1 tenth) = 5.0).

As we noted previously (see page 39), rounding is often used to find approximate solutions. For example:

- Which of these is the best approximation for 20.7 + 49.6?
  20 + 50 • 21 + 50 • 20 + 40 • 21 + 49
- Which of these is the best approximation for 79.5 − 40.2?
  80 − 50 • 800 − 400 • 80 − 40 • 79 − 40
- Which of these is the best approximation for 7.28 × 2.9?
  8 × 3 • 10 × 3 • 7 × 2 • 7 × 3
- Which of these is the best approximation for 8.9 ÷ 3.1?
  89 ÷ 31 • 10 ÷ 3 • 8 ÷ 3 • 9 ÷ 3

The answers to the above are, respectively:

21 + 50	giving an approximate answer of 71
80 − 40	giving an approximate answer of 40
7 × 3	giving an approximate answer of 21
9 ÷ 3	giving an approximate answer of 3

'Approximately' is a mathematical term, and needs to be identified as such.

• 'The England cricket team scored 198 runs in their first innings and 204 runs in their second innings. Approximately how many runs did England score?'

To find **'approximate'** answers, round the numbers in the question first and then do the 'easy' sum. So the number of runs is approximately equal to 200 + 200, which is 400 runs.

The symbol for 'approximately equal to' looks like this:

'≈'

So the above could be rewritten as:

Number of runs ≈ 400

Another example:

'Approximate:	(603 + 197)	÷ 41'
Approximate Answer:	(600 + 200)	÷ 40
	800	÷ 40
	20	

Using the symbol notation: (603 + 197)   ÷ 41 ≈ 20

Rounding can be used to help give quick and efficient 'answers'.

The Multiplication Square

A Multiplication Square is another prop that your children are likely to encounter in class, from the very early years up to and throughout secondary school. Similar in style to the **Hundred Square** mentioned earlier, the **Multiplication Square** is also often seen poster style on classroom walls. Your children are also likely to have individual copies in their book or tray.

The Multiplication Square is a very good aid to help children learn their times tables. It simply lists all the times tables up to 10 × 10 (or 12 × 12) in rows and columns . Children can be shown how to spot patterns and use the symmetry of the square to make learning their tables easier (see also pages 155 and 278). Again if you can get your own big and colourful **Multiplication Square** to stick up at home – great!

×	1	2	3	4	5	6	7	8	9	10
1	1	2	3	4	5	6	7	8	9	10
2	2	4	6	8	10	12	14	16	18	20
3	3	6	9	12	15	18	21	24	27	30
4	4	8	12	16	20	24	28	32	36	40
5	5	10	15	20	25	30	35	40	45	50
6	6	12	18	24	30	36	42	48	54	60
7	7	14	21	28	35	42	49	56	63	70
8	8	16	24	32	40	48	56	64	72	80
9	9	18	27	36	45	54	63	72	81	90
10	10	20	30	40	50	60	70	80	90	100

What is a multiple?

In simple terms, multiples are just your times tables facts.
So the multiples of 5 are: 5, 10, 15, 20, 25, 30, 35…

The sequence never ends as we can just keep adding another 5. So there are an endless, or infinite, number of multiples.

Another way of thinking about a multiple is to see it as a *result* of multiplying 2 whole numbers.

4 × 5 = 20   so 20 is a multiple of 4 and also a multiple of 5
3 × 6 = 18   so 18 is a multiple of 3 and also a multiple of 6

What is a factor?

Factor is another term your children will definitely need to know. Factors are the numbers you multiply together to make another number.

For example:

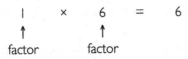

So 2 and 3 are factors of 6.

Similarly 1 and 6 are also factors of 6 because:

```
    1    ×    6    =    6
    ↑         ↑
 factor    factor
```

Prime numbers are an interesting set of numbers. A prime number has only two factors, 1 and itself. The number '1' is not prime as it has only one factor, namely '1' itself. However, '7' is a prime number as it has only two factors: 1 and 7.

But '8' is not prime as it has more than two factors, namely: 1, 2, 4 and 8.

The 'Sieve of Eratosthenes' is a typical exercise used with Year 7 pupils, to 'sieve' out prime numbers.

A **Hundred Number Square** is used:

1	2	3	4	5	6	7	8	9	10
11	12	13	14	15	16	17	18	19	20
21	22	23	24	25	26	27	28	29	30
31	32	33	34	35	36	37	38	39	40
41	42	43	44	45	46	47	48	49	50
51	52	53	54	55	56	57	58	59	60
61	62	63	64	65	66	67	68	69	70
71	72	73	74	75	76	77	78	79	80
81	82	83	84	85	86	87	88	89	90
91	92	93	94	95	96	97	98	99	100

The number 1 is not prime as it has only one factor, so 1 is crossed out.

The next number, 2, is prime as it has exactly two factors: 1 and itself. So we put a circle around 2 to mark it as prime. Now all the multiples of 2 are crossed out because they cannot, by definition, be prime (as they would all have 2 as a factor as well as 1 and themselves).

The next number is 3 and this is also circled as being a prime number. Now all the multiples of 3 are crossed out as they cannot be prime.

The next number that has not previously been crossed out is 5 and this is circled. All the multiples of 5 are then crossed out.

The next number not crossed out is 7 and this is the next prime number. All the multiples of 7 are now crossed out. Continuing in this fashion, the next prime number is 11. In this way we literally *sieve* out all the prime numbers up to 100:

2, 3, 5, 7, 11, 13, 17, 19, 23, 29, 31, 37, 41, 43, 47, 53, 59, 61, 67, 71, 73, 79, 83, 89, 97.

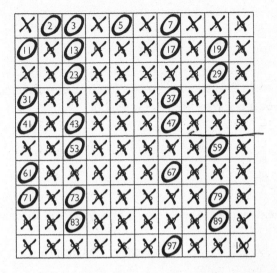

You may notice that apart from 2, all the prime numbers are odd. The reason that 2 is the only even prime number is because every other even number would have 2 as a factor.

ROMAN NUMERALS

The Roman number system – known as Roman numerals – was used all across Europe before being replaced by the Arabic number system that we use today some time in the Middle Ages. We can still see Roman numerals in use today but mainly for decorative purposes.

Roman numerals are based on 7 basic symbols. By combining these symbols the Romans could write any number they could conceive of. However, the Romans had not conceived of the idea of 0 and so had no need to write it!

I	=	1
V	=	5
X	=	10
L	=	50
C	=	100

D	=	500
M	=	1000

The system is to add the symbols together to make a number, until you are 1 away from a new symbol:

I	=	1
II	=	2
III	=	3

The next number (4) is one away from V (5), and so we write:

IV	=	4 meaning one before 5

And then carry on:

V	=	5
VI	=	6
VII	=	7
VIII	=	8

Now the number 9 is one before 10, and so we write:

IX	=	9 meaning one before 10

And then carry on again:

X	=	10
XI	=	11
XII	=	12
XIII	=	13
XIV	=	14
XV	=	15
XVI	=	16
XVII	=	17

XVIII	=	18
XIX	=	19
XX	=	20

30 is XXX but 40 is XL, as it is 10 away from 50. Forty-nine is XLIX.

60 is LX, 70 is LXX, 80 is LXXX but 90 is XC as it is 10 away from 100.

400 is CD as it is 100 away from 500. Six hundred is DC.

900 is CM as it is 100 away from 1000.

Quite often you can see the date on a bridge or building expressed in Roman numerals. This is the date on a bridge near me in the beautiful city of Bath:

MDCCCLXXXVI

and is:

1886 in Arabic numerals.

Children are often introduced to the Roman numerals partly as history, partly as geography (what we can see in the city in which we live) and partly as mathematics. It can be mathematically challenging to work out how to write the number '449' in Roman numerals (CDXLIX). And by demonstrating how difficult it is to do addition using the Roman numerals system – 2 numbers cannot be added using a simple algorithm of lining up one number under another – the beauty and ease of our more modern (Arabic) number system is highlighted.

The four operations of number

In maths texts you will often read about the **four operations of number**. Quite simply these are: **add, subtract, multiply** and **divide**. **Addition** and **subtraction** are like 2 halves of a whole: each will 'undo' or reverse the other. **Multiplication** and **division** operate in exactly the same fashion: each having the opposite effect to the other. The next four chapters are devoted to these **four operations of number**.

As a rule of thumb, the aim is for all pupils at the end of Year 6 to be equipped with basic mental, written and calculator methods – which they understand and can use correctly – to add, subtract, multiply and divide. In addition to this, in my experience, secondary school maths teachers are delighted if pupils arrive in Year 7 with a 'feel' for numbers and a real sense of wonder and excitement.

Understanding the connection of numbers, enjoying numbers, being confident to play and experiment with them – these things are far more important, in my opinion, than an arbitrary test result.

Now you are familiar with some fundamentals, feel free to explore the rest of this book in any order.

For clarity's sake, you will occasionally encounter some repetition throughout the following chapters – this means that you don't have to do too much page flicking to find out what things mean. From now on, each chapter can stand alone.

Enjoy!

2.
ADDITION

Here's the first of the **four operations of number**, introduced at the end of Chapter 1. Addition sounds simple and often is, but it's among the most important building blocks of our mathematical tower, so it is important to help young children feel confident in this fundamental area.

UNDERSTANDING THE BASICS
(RECEPTION AND YEARS 1 AND 2, AGES 5–7)

Children are likely to hear and use many terms that all relate to adding up: addition, add, sum, total, altogether, plus, counting on, and, equal, equals, makes, more, more than, addition table, increase, how many more to make…

As we touched on in the Introduction, your children will be expected to use lots of practical and mental methods in the very early years. This means that they will do a great deal of playing, counting, recognising, talking, singing and drawing. The aim will be for children to build their confidence with numbers.

I use the mathematical signs '+' (add) and '=' (equals) throughout this chapter in order to be concise. However, these signs are not introduced at school during the early stages. Children will learn about the associated words and understand the operation of addition before they are shown the symbols. To introduce the symbols any earlier can be counter-productive and simply too confusing for children to comprehend.

For very young children simply replace the symbols in the examples below with the words 'add' and 'equals'.

In the classroom, children will most likely be involved in plenty of 'maths chat' and it is very easy to do the same at home. Examples were given in the Introduction, such as 'If you have 4 sweets and I give you 1 (or 2 or 3…) more…'. Here are a few more to get you started:

- 'We have 10 bricks, how many more do we need to make that tower?'
- 'If you have 6 sweets and Connor has 6 sweets, how many sweets are there altogether?'
- 'How much older than you is your cousin Lizzie?'
- 'Look at the digits on that car number plate. What do they add up to?'
- 'What do all the numbers on a dice add up to?'
- 'Let's add together the digits in our phone number.'

You'll be able to think of ones your children will like. And it will help if you can keep looking out for opportunities to do this kind of exercise together more (not less) often as your children get older.

The Number Line is a very simple and effective tool to help with adding up. It enables children to visualise numbers as a sequence – which can help with counting on. (Children will be able to 'see' that addition and counting on are related.)

Initially children will be shown how to move their fingers along it to count on. For example, if children are being asked to 'do 4 add 2', they should start by putting their finger on number 4 and then *jump* or move their finger *2 spaces* to the *right*, ending up at the answer 6.

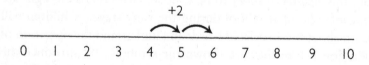

You will notice that I said children should move or *jump* their fingers 2 '*spaces*' and not digits. It's very important to encourage children to count the 'spaces' or 'jumps' between the numbers, rather than the digits themselves. A very common misunderstanding for young children (and sometimes the not-so-young) is that they count the numbers they see.

Here is an example:

- 'Billy has 5 red beads. Cassie passes him another 3 red beads. How many red beads does Billy have now?'

To answer this question correctly, Billy should put his finger on the 5 and *jump* 3 spaces to the right. Billy now has 8 red beads.

But some children will look at this bit of the Number Line

```
......_____........
    5       6       7
```

and give the *incorrect* answer of 7. They have counted the numbers instead of the *spaces* (or gaps or jumps) *between* the numbers and arrived at the wrong answer.

The Number Line can, of course, be extended as far as needed. Have a look at the line below and imagine your children being asked to 'do 8 add 5', or '2 add 10'.

```
0 1 2 3 4 5 6 7 8 9 10 11 12 13 14 15 16 17 18 19 20
```

And Number Lines can be used to add more than 2 digits. For example: 3 + 2 + 7

```
   +2          +7
```
```
0 1 2 3 4 5 6 7 8 9 10 11 12 13 14 15 16 17 18 19 20
```

Over time and with experience children can use Number Lines that don't start at 0. If we ask children a question along the lines of, '*If you have 27 strawberries, and I give you another 8, how many will you have?*' they can use a section of the Number Line starting where they need it to:

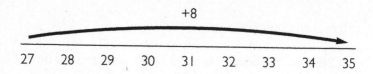

Later the Number Line will become even more abstract and used as a prompt for children to do their calculations (see page 86).

Complements

In early years maths we often hear of **complements to 5** and **complements to 10**. A complement of something is simply what is required to complete it. For example:

- **'Complements to 5'** simply means: 'What more is needed to make a total of 5?'

For example: If there are 5 teddy bears and 3 of them have a plate, how many more plates do we need so that all the teddy bears have one for their picnic?
Answer: 2

Children often use their fingers on one hand to help them with this, which is an excellent visual starting point. Children can easily see the links between the facts as they look at their fingers and try to move them apart: 1 + 4 = 5; 2 + 3 = 5 and so on.

Of course, in time children will need to become competent with their complements without relying on fingers.

- **Complements to 10** similarly means: 'What more is needed to make a total of 10?'

For example: 'Oliver is allowed to choose 10 friends to come to his birthday party. He has already chosen 7. How many more friends can he invite?'
Answer: 3

Again fingers can be used in the early days – in this case, of both hands – but, again, in time children will need to move beyond this. Being able to recognise and recall complements to 10 with ease is an essential building block for basic addition.

Addition number bonds

These may sound complicated but really they are not. All this term means is how two digits 'bond' or connect to each other.

Addition number bonds (or **addition facts**) are pairs of numbers added together. For example, all the addition number bonds for 7 are listed below:

$$0 + 7 \bullet 1 + 6 \bullet 2 + 5 \bullet 3 + 4 \bullet 4 + 3$$
$$5 + 2 \bullet 6 + 1 \bullet 7 + 0$$

Likewise for 9:

$$0 + 9 \bullet 1 + 8 \bullet 2 + 7 \bullet 3 + 6 \bullet 4 + 5$$
$$5 + 4 \bullet 6 + 3 \bullet 7 + 2 \bullet 8 + 1 \bullet 9 + 0$$

Simple number bonds are very helpful to know by heart, so basic addition can happen easily.

It is especially important for children to practise and use the number bonds to 10 and to 20 regularly. Looking for pairs of numbers that add up to either 10 or 20 can really speed things up! I always think of it as putting 2 pieces of a jigsaw together.

Here are some examples to show you what I mean:

$$4 + 6 + 8 \rightarrow 10 + 8 = 18$$

$$9 + 3 + 17 \rightarrow 9 + 20 = 29$$

$$8 + 13 + 2 \rightarrow 10 + 13 = 23$$

$$19 + 14 + 1 \rightarrow 20 + 14 = 34$$

Most children will probably be introduced to the '+' (add) and '=' (equals) signs some time in their first or second year at school. Children may start relating their 'maths chat' problems to add sums, using these symbols. This is often referred to as writing 'number sentences'.

Number sentences (or **maths sentences**) may be an unfamiliar term to you, but it's actually a very simple way of starting to write down, or record, mental calculations. **Addition number sentences** use the '+' and '=' symbols.

Imagine Javaid had been doing some adding sums. He started with 3 blue beads, then chose 4 yellow beads and then 2 red beads. He works out how many beads he has altogether. He adds them all up and tells the teacher he has 9 beads altogether. The teacher is very pleased and then asks Javaid to record this as a number sentence.

Javaid writes:

3 beads + 4 beads + 2 beads = 9 beads

Or he might just write:

3 + 4 + 2 = 9

And that's all there is to it.

You will probably have seen some very simple examples of number sentences in children's activity books or on the back of a cereal box.

Using symbols: Symbols such as ■ and ▲ can be used to stand for numbers. This isn't tricky stuff really, but can be confusing if you're not expecting them.

For example:

4 + 5 = ■	Answer: ■ = 9
3 + ▲ = 7	Answer: ▲ = 4
⬟ + 4 = 9	Answer: ⬟ = 5

By the end of Year 2 children will probably be familiar with: **multiples of 10 that total 100**. This idea is easiest to explain with examples:

40 + ? = 100	Answer: ? = 60
70 + ? = 100	Answer: ? = 30

Often symbols are used:

20 + ■ = 100	Answer: ■ = 80
▲ + 10 = 100	Answer: ▲ = 90
⬟ + 40 = 100	Answer: ⬟ = 60

So this is a very similar idea to the number bonds we saw earlier, but instead of a list of every pair of numbers that total 100 (which would be rather long), it focuses just on these:

10 + 90 = 100
20 + 80 = 100
30 + 70 = 100
40 + 60 = 100
50 + 50 = 100
60 + 40 = 100
70 + 30 = 100
80 + 20 = 100
90 + 10 = 100

Again, it is really useful for children to know these off by heart, perhaps by the end of Year 2.

Adding 0 ('zero' or 'nought') *always* leaves a number unchanged. This seems a rather abstract idea until your children understand that adding nothing does nothing:

6 + 0 = 6

This may seem obvious – but it isn't. Examples always help children's understanding: If Lewis had 6 Easter eggs and then was given no more, he would still have 6 Easter eggs.

6 Easter eggs + 0 Easter eggs = 6 Easter eggs

Addition doubles (or pairs) are simply doubles of a number. For example:

1 + 1 = 2
2 + 2 = 4
3 + 3 = 6
4 + 4 = 8
5 + 5 = 10
6 + 6 = 12   ...and so on

Children will initially learn to double numbers up to 5, then up to 10. By the end of Year 2 children will have regularly practised doubling numbers up to 20 + 20 = 40.

Again, it will help your children if they can work these sums out quickly, and remember some by heart.

Holding up fingers on hands can be a really good way to introduce doubling. For example, one finger on each hand for 1 + 1, two fingers on each hand for 2 + 2, and so on up to 5 + 5. This will encourage them to use their fingers to keep checking and counting.

It's worth mentioning at this stage that it is perfectly normal and necessary for children to keep checking things out. Just because they've worked out that if they hold up all their fingers on both hands they've got 5 fingers and 5 fingers, which make 10 fingers, they won't **know** 5 + 5 = 10 as a fact. They've got to keep checking.

Over time they'll realise it is always the case and begin to understand that 5 + 5 is *always* 10 (and that 1 + 1 = 2 and 2 + 2 = 4 and so on).

A classic example your children will probably be shown is how to do sums like 19 + 19 very quickly. If your children know by heart that 20 + 20 = 40, they can use this fact to help them calculate other doubles quickly:

$$19 + 19 = 20 + 20 - 1 - 1$$
$$= 40 - 2$$
$$= 38$$

$$21 + 21 = 20 + 20 + 1 + 1$$
$$= 40 + 2$$
$$= 42$$

Similarly **near-doubles** can be calculated quickly:

$$6 + 7 = 6 + 6 + 1$$
$$= 12 + 1$$
$$= 13$$

$$15 + 16 = 15 + 15 + 1$$
$$= 30 + 1$$
$$= 31$$

Today's children are repeatedly reminded and encouraged to use mental strategies for adding. One such strategy is always to count on from the greatest number.

For example, given the sum 2 + 7, your children might be taught to 'put' the 7 in their head and count on 2.
Answer: 9

For this to work we need to appreciate that **addition is commutative**. Your children won't need to learn this word, but it can be a helpful one for adults to know. It simply means that:

2 + 7 is the same as 7 + 2

This principle is important because often children will find the second sum much easier than the first. Adding on 2 just seems a simpler task. (It is easier to visualise counting on 2.) So we can let them know that it doesn't matter: with adding up, it is fine to do the 'easier' sum. Try it yourself with bigger numbers but remember, in your head only. Which seems 'easier': 49 + 106 or 106 + 49?

When adding slightly bigger numbers such as 23 + 4, children will still be encouraged to use the mental method as above, or a Number Line.

When children progress to adding two 2-digit numbers (such as 34 + 23) they will be taught to partition the numbers first. But before children can be taught about partitioning, they must have a secure knowledge of **place value**, introduced in Chapter 1 (see pages 16 and 30).

Partitioning

Partitioning may be another complex-sounding idea – but again it isn't! Partitioning simply means to split a number up into bits. Examples will show you what I mean.

- 23 can be partitioned into → 20 + 3
- 37 can be partitioned into → 30 + 7
- 19 can be partitioned into → 10 + 9
- …and that's it! Easy-peasy!

All the above examples have been split into **tens** and **units**.
Bigger numbers can be similarly split into **hundreds**, **tens** and **units**, for example:

- 358 can be partitioned into → 300 + 50 + 8

WAYS OF PARTITIONING AND RECOMBINING

Partitioning numbers into their respective hundreds, tens and units is a common way of partitioning, but it is not the only way. For example, to do 45 + 17 in our heads, we may choose to partition the 17 like this:

> 45 plus 17
> 45 plus (15 + 2)
> (45 plus 15) + 2
> 60 + 2

Answer: 62

For young children partitioning even very small numbers can help make a sum simpler to do. For example: 6, 7, 8 or 9 can all be thought of as '5 and a bit'.

> 6 + 8   can be thought of as
> (5 + 1) + (5 + 3)   which can then be thought of as
> 5 + 5 + 1 + 3

Now the 5 + 5 is easy, owing to all our practice with number bonds to 10 and also doubles of numbers. So, now we recombine (that is, put the partitioned bits back together) and it becomes:

> 5 + 5 + 1 + 3   which is the same as
> 10 + 4   giving the answer
> 14

There are always different ways to get to the right answer. Children may choose to partition in different ways and that's fine. The last example could also look like this:

> 6 + 8   can be thought of as
> 6 + (6 + 2)
> 12 + 2
> 14

To use partitioning to add together two 2-digit numbers, the second number is partitioned. For example:

$$34 + 23 \rightarrow 34 + (20 + 3) \rightarrow 54 + 3 = 57$$

$$71 + 19 \rightarrow 71 + (10 + 9) \rightarrow 81 + 9 = 90$$

Later (in Year 2 or Year 3) both numbers will be partitioned before being added or recombined. This is in preparation for the early written methods of addition (explored in the next section – see page 85). For example:

$$47 + 27 \rightarrow 40 + 7 + 20 + 7 \rightarrow 60 + 14 = 74$$

$$23 + 38 \rightarrow 20 + 3 + 30 + 8 \rightarrow 50 + 11 = 61$$

The Hundred Square

A **Hundred Square**, as introduced in Chapter 1 (see page 18), is simply a grid of the first 100 counting numbers arranged in rows as below, and is another tool for helping with addition:

1	2	3	4	5	6	7	8	9	10
11	12	13	14	15	16	17	18	19	20
21	22	23	24	25	26	27	28	29	30
31	32	33	34	35	36	37	38	39	40
41	42	43	44	45	46	47	48	49	50
51	52	53	54	55	56	57	58	59	60
61	62	63	64	65	66	67	68	69	70
71	72	73	74	75	76	77	78	79	80
81	82	83	84	85	86	87	88	89	90
91	92	93	94	95	96	97	98	99	100

At first your children may be asked, '*If we start with any number, what happens when we add on 10?*' Very quickly they will 'see' that they just need to look down the column.

Spotting patterns such as:

$$36 + 10 = 46$$
and $$46 + 10 = 56$$
and $$56 + 10 = 66$$
and $$66 + 10 = 76$$
and $$76 + 10 = 86$$

helps with understanding number and place value, as children can see the number of **tens** changing whilst the **units** remain the same (see also page 18).

It is certainly worth practising adding 10 to any 2-digit number, for example:

$$34 + 10 = 44$$
$$82 + 10 = 92$$
$$67 + 10 = 77$$
$$75 + 10 = 85$$

When adding with a Hundred Square it is often easier to start with the biggest number. For example, in the case of $13 + 54$ your child will be shown to start with 54 and then add on 13. Initially children will probably just count on 13 squares and arrive at the answer 67. But by the age of 7 or so your children may be partitioning numbers first.

For example:

$$38 + 54$$

Your children will be encouraged to start with the biggest number and find this on the Hundred Square. So they place their finger on 54. Then they will partition the other number into **tens** and **units** ($38 \rightarrow 30 + 8$).

To add the tens (30) they move their finger down the columns.

In this case they move down 3 columns because there are 3 tens (30). So their finger is now on 84 (54 + 30 ➜ 84). Then, to add the units they move their finger along the row to the right, counting on 8 spaces to arrive at the answer: 92.

1	2	3	4	5	6	7	8	9	10
11	12	13	14	15	16	17	18	19	20
21	22	23	24	25	26	27	28	29	30
31	32	33	34	35	36	37	38	39	40
41	42	43	44	45	46	47	48	49	50
51	52	53	(54)	55	56	57	58	59	60
61	62	63	64	65	66	67	68	69	70
71	72	73	74	75	76	77	78	79	80
81	82	83	(84)	85	86	87	88	89	90
91	(92)	93	94	95	96	97	98	99	100

In summary: to add the **tens** travel *down* the columns. To add the **units** travel to the *right* along the rows.

When using the Hundred Square, your children may 'see' or discover easier ways to do things. For example, an easy way to add on 9 would be to add on 10 and then take off 1. Similarly to add on 11 it is easy to add on 10, then another 1. These are examples of **adjusting**.

Adjusting

Adjusting is about making things easier, and once more it's the sort of calculation we do in our heads all the time.

To add 19 to any number we could start by adding 20 and then **adjusting** the answer by subtracting 1. For example:

```
     16 + 19
Do 16 + 20
and then adjust by taking away 1
     36 − 1 gives us
     35
```

Add 21 to any number by adding 20 and then **adjusting** the answer by adding 1.

For example:

```
     16 + 21
Do 16 + 20
and then adjust by adding 1
     36 + 1
     37
```

Obviously this can be extended to bigger numbers and greater adjustments.

For example:

```
     253 + 102
Do 253 + 100 and then adjust by adding 2
     374 + 97
Do 374 + 100 and adjust by taking away 3
```

The **opposite operation** to addition is **subtraction,** and this can be really useful to know if we accidentally add on too much. In other words subtraction **reverses** or **undoes** addition. (Technically speaking, subtraction is the **inverse** of addition.)

Oliver and his baby brother Matthew are both given 8 sweets. Matthew happily passes 3 of his sweets to Oliver. Oliver now has 11 sweets.

```
8 + 3 = 11
```

Their mum insists Oliver return the sweets Matthew has given him. So Oliver reluctantly takes away 3 sweets from his pile, knowing he'll be back to just the 8 sweets he started with.

11 − 3 must equal 8

This rule is very useful if children accidentally add on too much, which they might do when using a calculator. To correct their mistake, they can just subtract what they added on and they are back to where they started.

If this all sounds obvious, it may not be to your children. I've met many secondary-age pupils who do not realise this, or don't understand it sufficiently to be able to use it to good effect.

An **addition table** is a typical exercise your children might be given. An addition table is nothing more than a grid of numbers your children may be asked to add together, either in a sequential order or alternatively all jumbled up. An example of each is shown below (both partially completed).

It's just another way of helping children practise addition and another useful thing for you to know.

+	1	2	3	4	5
1	2	3	4	5	6
2	3	4	5	6	7
3	4	5	....		
4	....				
5	....				

+	5	2	1	6	4
2	7	4	3	8	6
8	13	10	9	....	
3	....				
1	....				
7	....				

A **Magic Square** is another exercise that children might encounter to help them practise adding up. It is fun and most children like doing it. A Magic Square is a square in which the numbers in every row, column and diagonal add up to the same total. For example:

6	1	8
7	5	3
2	9	4

In this example the total for each row, column and diagonal is 15.

Magic Squares can be used at many different levels and that is why they can be a useful exercise from primary level well into the later years at secondary school. At the most basic level, children can be given a square and asked to say whether it is magic or not – simply by adding up all the rows, columns and diagonals. Later they may be given a square with a few numbers missing and be asked to fill in the numbers to make it magic. For example:

8	*	4		**Answer:** 8	3	4
1	5	*		1	5	9
*	7	2		6	7	2

Even later they can be given just the total and asked to construct a Magic Square.

Magic Squares can also be of different sizes. The bigger they are, the harder they get. For example:

1	15	14	4
12	6	7	9
8	10	11	5
13	3	2	16

The total for each row, column and diagonal in this example is 34.

All the examples of magic squares above have used consecutive numbers.

(**Consecutive numbers** are simply counting numbers and digits that come after one another, such as 1, 2, 3, 4... or 6, 7, 8... or 23, 24, 25, 26... or 235, 236, 237...)

Magic Squares can also use digits and numbers that are *not* consecutive.

For example:

9	5	10
9	8	7
6	11	7

The total for each row, column and diagonal is 24.

MAGIC SQUARES AND ALGEBRA

When children reach secondary school they may be shown how to use algebra to find missing numbers in a magic square. Here is a very neat example:

*	2	13
12	*	*
*	14	7

We can rewrite the square replacing two of the missing numbers with letters:

x	2	13
12	*	*
y	14	7

Now, as we know all the rows and columns add up to the same amount, we know that the sum of the top row is the same as the sum of the first column. In this example:

$$x + 2 + 13 = x + 12 + y$$

We can rewrite this as $\quad x + 15 = x + 12 + y$

Now subtract x from both sides $\quad 15 = 12 + y$

Now subtract 12 from both sides $\quad 3 = y$

Now we can rewrite the square, replacing y with 3:

x	2	13
12	*	*
3	14	7

And now from looking at the bottom row we can find the total. The total is 24, so we can now put in all the missing numbers:

9	2	13
12	8	4
3	14	7

Very neat, I think – and a good party trick!

Related addition facts

Using **related addition facts** simply means using things we know to help us solve things we don't know!!

For example, to calculate 40 + 80, children may use the fact that they *know* 4 + 8 = 12, along with their knowledge of place value, to derive that 40 + 80 = 120.

Spotting and looking for patterns also helps, as our brains naturally work this way. We just can't help spotting and remembering patterns, which is why a neatly patterned telephone number is eagerly sought after by taxi firms. Playing to this natural strength can be really helpful for children of all ages, and adults too.

For example:

	4 +		8	=	12
Similarly	40 +		80	=	120
and	400 +		800	=	1200
and	4000 +		8000	=	12000

Another example of using related addition facts is using the **complements to 10** to help with calculations for larger sums.

One example might be 27 + 43. Your children might know how to piece together the 7 and the 3 like a jigsaw to come up with the complement of 10. Then it's just a matter of adding 20, 40 and 10.
Answer: 70

Here are some more examples:

$$54 + 26 = 80$$
$$25 + 65 = 90$$
$$51 + 19 = 70$$
$$13 + 87 = 100$$
$$62 + 18 = 80$$

Adding 100 is something children will practise a lot from Years 2 or 3 onwards.

The process of adding 100 to any 3-digit number is one that it's certainly worth practising again and again. At this stage, it might be best to keep it manageable by avoiding crossing into the thousands. For example:

$$683 + 100 = 783$$
$$837 + 100 = 937$$
$$219 + 100 = 319$$
$$750 + 100 = 850 \quad \text{...and so on}$$

The **associative law** is another way of making 'adding up' easier. Once again you don't need to be familiar with the terminology – just know what it means. And what it means is this...

We saw earlier that: $2 + 7 = 7 + 2$ (that is, addition is **commutative**). Now we can take this idea a stage further.

Looking at the sum:

$35 + 28 + 15$ we can see that it is exactly the same as:
$35 + 15 + 28$, which is the same as
$15 + 28 + 35$, which is the same as
$15 + 35 + 28$, which is the same as
$28 + 15 + 35$, which is the same as
$28 + 35 + 15$

In other words, it doesn't matter in what order you add up the numbers. This is the **associative law**.

This is significant because it helps us make a calculation as simple as possible.

So, for the sum

$$35 + 28 + 15$$

it might be easier to do

$$35 + 15 \text{ first, which equals}$$
$$50$$

then add the 28

$$50 + 28 \text{ equals}$$
$$78$$

Again, if this is the sort of thing you do in your head anyway without really thinking about it, bear in mind that children usually need it to be pointed out to them.

MOVING FORWARD
(YEARS 3 AND 4, AGES 7–9)

We've covered the basics in addition above, and this is all still used as your children progress, generally using bigger numbers. The **Hundred Square** and **Number Line** will also continue to be very much in evidence.

Now it can be satisfying to see our children moving towards what's known in the trade as '**pencil-and-paper**' methods. But please bear in mind that initially these written methods are informal and used to support and extend what your children have already been learning to do. Writing things down should simply build on their understanding so far.

Mental maths is still very important. Children will secure and practise their mental methods alongside learning new written methods.

The written methods are built up in stages, with the intention of enabling children to progress to the next stage when they feel confident to do so. The end point is the 'standard column method' with which you may be familiar from your own school days.

For children to use these written methods with ease, they must be able to build on everything they have been taught so far and have a very secure understanding of **place value** (see also pages 16 and 30).

In summary: to add with ease children need to be able to:
- Recall all their **number bonds** up to 9 + 9.
- Know **complements** to 10.
- Add 1-digit numbers in their head, such as 3 + 9 + 5.
- Add multiples of 10 (such as 40 + 80) by using the **related addition fact** that 4 + 8 = 12, and their knowledge of **place value**.
- Similarly add multiples of 100 (such as 400 + 800).
- **Partition** 2-digit and 3-digit numbers in different ways.

The stages...

Stage 1: The Empty Number Line

At this point, the number line is still used, but it becomes a little more abstract. Numbers need not be written sequentially. In fact, only a sketch of a line with a few jottings may be all that is recorded. This is referred to as an **Empty Number Line**.

Children will probably be familiar with partitioning numbers into hundreds, tens and units. Now they will partition in other ways too. The Empty Number Line is a way of recording the steps on the way to reaching the final answer.

For example, for the sum 18 + 7, children may 'jump' from 18 to the nearest **ten** (20 in this case) and then see they have a further 5 to add on.

So the Number Line might look something like this:

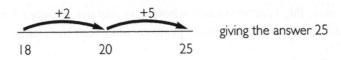

giving the answer 25

18   20   25

For the sum **74 + 63**, children may partition the second number into **tens** and **units** and sketch a Number Line something like this:

giving the answer 137

The 'scale' of the Number Line is now irrelevant. The line is simply used as a prop to aid calculation. Another example: **27 + 48** might look like this:

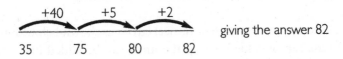

giving the answer 75

Or – as children are often taught to start with the bigger number when adding – this sum might also be shown like this:

giving the answer 75

In both of the cases above, the second number has been partitioned into **tens** and **units** and then the **units** split up again, so as to *jump* to the nearest **ten** (70 in this case).

Another example for **35 + 47** might look like this:

giving the answer 82

It's all about making the addition as easy as possible. The Empty Number Line can be a quick and easy prompt to aid addition for all ages. Pupils at secondary school may *still* use Number Lines, similar to these, to help them perform and check their sums.

Stage 2: Partitioning

The next stage is to use partitioning in a way that links mental methods and written methods. Children will be encouraged to start *writing down* their partitionings (using **hundreds**, **tens** and **units**) and *showing* their workings. This will resemble simple jottings and is used to help retain information.

For example, for the sum 45 + 89, the 45 will be split into 40 and 5 and the 89 into 80 and 9. This can then be *written down* as:

$$45 + 89 \rightarrow \quad 40 + 80 + 5 + 9 \rightarrow \quad 120 + 14 = 134$$

So the **tens** are added together and then the **units**, to form what are known as **partial sums**. These partial sums are then added together to give the final answer.

Then children will be encouraged to write the partitioned numbers underneath one another, and then add up as before:

$$
\begin{array}{ll}
45 + 89 \text{ may be written as } 45 = & 40 + 5 \\
+\,89 = & 80 + 9 \\
\hline
& 120 + 14 = 134
\end{array}
$$

These mark the very early stages of written sums.

Stage 3: Expanded method in columns

The next step moves on to a vertical layout as shown below. Either the **tens** can be added first *or* the **units** can be added first – it is worth showing your children that the answer is the same whichever way they do it.

For example: 52 + 37 might look like this:

```
        52          or                    52
        37                                37
        ──                                ──
50 + 30 →   80              2 +  7 →       9
 2 +  7 →    9             50 + 30 →      80
            ──                            ──
            89                            89
```

Adding **tens** first   Adding **units** first

With these informal methods of recording sums, it doesn't matter which version your children use. It might be more natural for them to add the **tens** digits first, as it's probably what they've been used to doing using mental methods. However, as time progresses they *should* be encouraged to add the **units** first as this will be a step towards the more formal standard written methods. In all the examples below I have only shown the version that adds the **units** first, to avoid too much repetition.

For these methods to work, your children must know that **units** line up under **units**, **tens** under **tens** and so on – just as we always did at school. Another example:

```
47 + 86 may now look like this:      47
                                     86
                                     ──
                      7 +  6 →       13
                     40 + 80 →      120
                                    ───
                                    133
```

This expanded method will lead children on to the more efficient compact column method outlined in detail later (see page 93).

Try adding 51 and 27. Do it in your head first. Now write it down as you were taught at school. In your head you probably added the *tens* first (and then added on the 1 and the 7). Writing it down you probably added the units first. It is worth remembering that – up to this point – children have been doing all their sums in their head. So adding the *tens* first is probably much more natural for them at this stage.

When your children first start out on these written methods, early examples would ensure that *neither* the **tens** boundary nor **hundreds** boundary were crossed. Then, as their competency and confidence grow, either the tens **or** hundreds boundary will be crossed, and later both.

WHAT DOES 'CROSSING THE TENS BOUNDARY' MEAN?

This isn't a novel method of scoring cricket, but to explain what it is, it's easier to show you:

$4 + 3 = 7$ Here we started with **units** only and the answer is still **units** only – so no boundary has been crossed.

$4 + 8 = 12$ Here the answer is now **tens** and **units** so the **tens** boundary has been crossed.

$27 + 2 = 29$ Here we started with 2 **tens** and some **units** in the sum and the answer is still 2 **tens** and some **units**, so no new boundary has been crossed.

$27 + 4 = 31$ When adding together the 7 **units** and the 4 **units** a **tens** boundary has been crossed.

$64 + 20 = 84$ Here no boundary has been crossed.

$64 + 28 = 92$ When adding together the 4 **units** and the 8 **units** a **tens** boundary has been crossed.

$90 + 30 = 120$ We started with **tens** and **units** in the sum and the answer is **hundreds**, **tens** and **units** so the **hundreds** boundary has been crossed.

$95 + 28 = 123$ When adding together the 5 **units** and the 8 **units** a **tens** boundary has been crossed and when adding together the 90 and the 20 a **hundreds** boundary has been crossed.

For example:

64 + 23	64	
	23	
	—	
4 + 3 →	7	
60 + 20 →	80	
	—	
	87	neither boundary crossed

58 + 34	58	
	34	
	—	
8 + 4 →	12	**tens** boundary crossed
50 + 30 →	80	
	—	
	92	

72 + 65	72	
	65	
	—	
2 + 5 →	7	
70 + 60 →	130	**hundreds** boundary crossed
	—	
	137	

86 + 45	86	
	45	
	—	
6 + 5 →	11	both **tens** and
80 + 40 →	120	**hundreds** boundaries crossed
	—	
	131	

Once these techniques have been mastered, children are asked to tackle bigger numbers that require an extra line in the sum.
    For example:

342 + 37	342	
	37	
	—	
2 + 7 →	9	
40 + 30 →	70	
plus 300	300	
	—	
	379	no boundary crossed

456 + 38	456	
	38	
	—	
6 + 8 →	14	**tens** boundary crossed
50 + 30 →	80	
plus 400	400	
	—	
	494	

572 + 56	572	
	56	
	—	
2 + 6 →	8	
70 + 50 →	120	**hundreds** boundary crossed
plus 500	500	
	—	
	628	

686 + 59	686	
	59	
	—	
6 + 9 →	15	both **tens** and
80 + 50 →	130	**hundreds** boundaries crossed
plus 600	600	
	—	
	745	

Once children are secure and confident with this method of addition, the more formal column method can be introduced. This *may* be in Years 4, 5, 6 or later.

Stage 4: Column method (or compact standard method)

Ah, a familiar bit! Just like we did at school. This is adding **units**, **tens** and **hundreds** in columns using 'carrying'. Bear in mind that if we need to 'carry' then we know a boundary has been crossed.

The logic and process remains exactly the same as for the expanded method outlined above. It is simply a more compact and concise way of recording.

For example, for 58 + 34 Adam would write it out as below and say, '*8 add 4 is 12 so I'll carry the 10 and write down the 2. And 50 add 30 plus the 10 I carried, equals 90. So the answer is 92.*'

```
58 + 34      58
          +  34
          ——
             92
          ——
          1
```

                    tens boundary crossed (that is, 'carry 10')

You will note that 'carry' digits are recorded below the line. We now use the words '*carry 10*' or '*carry 100*' (not '*carry 1*' as we might have done at school).

Here are some more examples:

```
425
 37
———
462
———
  1
```

                    tens boundary crossed (that is, 'carrry 10')

```
    362              576
     83               58
    ---              ---
    445              634
    ---              ---
     I                I I
```

hundreds boundary crossed (that is, 'carry 100')

both boundaries crossed (that is, 'carry 10' then 'carry 100')

Children *must* check that they have arranged **units** under **units**, **tens** under **tens** and so on in order to get the right result.

This column addition method is an efficient and reliable one and can be used with whole numbers and decimals (see pages 96 and 98 for more advanced examples).

Is it reasonable?

Talking of checking, it is a really good idea to get your children into the habit of checking whether their answer is reasonable. By that, I mean applying a '*does it make sense?*' test to the answer.

For example, 132 + 45 cannot equal 77. Why? Because 77 is smaller than one of the numbers we started with.

Checking that answers broadly 'make sense' is a basic but valuable idea (for adults as well as children!) but one which, in my experience, is often under-used and under-rated.

TO THE TOP
(YEARS 5 AND 6 PLUS, AGES 9–12 PLUS)

As your children continue on their mathematical journey, all the rules and methods for addition we've covered in the sections above are still in daily use. Expect bigger numbers and more complex questions, but the same principles are at work. In this section, we familiarise ourselves with some more terms and methods, and make the move into decimal numbers and negative (minus) numbers.

Compensation is another method of simplifying addition: basically, it just means adding too much and then taking some off.

For example, in the sum '346 + 289' the 289 is initially 'replaced' with 300 as this is an easier number to 'handle'. Then, because we have added too much we need to compensate by taking some off, or subtracting.

```
        346
  +     289
        ———
        646    (346 + 300)
  −      11    because we had added 11 too many
        ———    (the difference between 300 and 289 being 11)
        635
```

This can be illustrated with a Number Line:

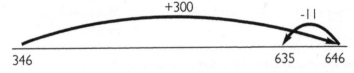

And another example:

```
       6764
  +    1750
       ————
       8764   (6764 + 2000)
  −     250   because we had added 250 too many
       ————   (the difference between 2000 and 1750
       8514   being 250)
```

And with a Number Line:

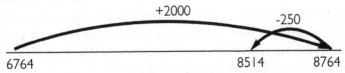

The column (or compact standard) written methods using 'carrying'

By about age 9, most children will have been introduced to this method (and so it is explained in more detail in the previous section) but they will probably start using it more fluently in Years 5 and 6. As they get older and grow more confident in applying it, the numbers used in the sums will usually get much bigger (or much smaller!). Here are some examples:

47	352	366
36	87	458
83	439	824
I	I	I I

tens boundary crossed (that is, 'carry 10')	**hundreds** boundary crossed (that is, 'carry 100')	**tens** and **hundreds** boundaries crossed (that is, 'carry 10' then 'carry 100')

And some examples with bigger numbers:

1272	2557	7865
349	784	4287
1621	3341	12152
I I	I I I	I I I I

tens and **hundreds** boundaries crossed	**tens, hundreds** and **thousands** boundaries crossed	**tens, hundreds, thousands** and **tens of thousands** boundaries crossed

Children must continue to remember to arrange **units** under **units**, **tens** under **tens** and so on in order to get the right result. This is even more significant when they are asked to add several numbers with different numbers of digits:

```
 5631
   54
  802
    9
 2373
 ____
 8869
 ____
 1 1 1
```

Adding with decimals

If you are coming to this again for the first time since school, you may like to refer to the 'Moving Forward' sections of Chapters 1 and 7 (see pages 34 and 315) to refresh your memory about decimal numbers before you read this bit.

A good way to introduce the idea of adding decimals may be to start with money:

> 25p + 55p + 12p
> can be also be shown as £0.25 + £0.55 + £0.12

As we saw earlier, children will first be expected to add up in their head (mental methods) before attempting the standard written methods.

The column (or compact standard) written method for adding decimals is just the same as we've seen above. Children just need to know and remember a few extra things:

- The decimal points must always line up underneath each other.
- Exactly as the **hundreds, tens** and **units** need to line up in straight columns, so do the **tenths** and **hundredths** (and **thousandths** and so on).

So the above sum would look like this:

```
0.25
0.55
0.12
────
0.92
────
 1
```

It is particularly important to remember this when adding up amounts with different numbers of decimal places; for example:

$$342.8 + 18.25 + 0.32$$

should look like this:

```
342.8
 18.25
  0.32
──────
361.37
──────
 1 1
```

Then there are sums with different units such as £4.74 + 63p or 13.5 kilograms + 250 grams.

Before we do anything else, we must convert the amounts into the same standard units, so 63p becomes £0.63 and 250 grams becomes 0.250 kilograms (because there are 1000 grams in a kilogram).

```
4.74                13.5
0.63                 0.250
────                ──────
5.37                13.750
────                ──────
 1
```

Adjusting

The same method of simplifying a sum explained in 'Understanding the Basics' (adding on a similar, but more convenient number and then adjusting the result up or down – see page 78) works just the same with bigger numbers and with decimal numbers.

For example, add 9, 19, 29, 39... or 11, 21, 31, 41... to any number by adding 10, 20, 30, 40... respectively and adjusting by 1. This method makes it hugely easier to do these sums in our heads.

56 + 19	add 20 and then subtract 1
62 + 31	add 30 and then add 1
356 + 39	add 40 and then subtract 1
673 + 81	add 80 and then add 1
1256 + 99	add 100 and then subtract 1
4583 + 201	add 200 and then add 1
3256 + 1999	add 2000 and then subtract 1
8739 + 4001	add 4000 and then add 1

Once the idea of adjusting has been grasped, children will be happy to adjust by more than one.

82 + 23	add 20 and then add 3
67 + 18	add 20 and then subtract 2
374 + 96	add 100 and then subtract 4
2025 + 1995	add 2000 and then subtract 5
6573 + 2006	add 2000 and then add 6

Similarly for decimal numbers add 0.9, 1.9, 2.9, 3.9... or 1.1, 2.1, 3.1, 4.1... to any number by adding 1, 2, 3, 4... respectively and adjusting by 0.1.

2.6 + 0.9	add 1 and then subtract 0.1	**Answer:** 3.5
8.7 + 4.9	add 5 and then subtract 0.1	**Answer:** 13.6
12.3 + 7.1	add 7 and then add 0.1	**Answer:** 19.4
32.9 + 2.1	add 2 and then add 0.1	**Answer:** 35.0
78.6 + 3.1	add 3 and then add 0.1	**Answer:** 81.7

Again, children can learn to adjust by greater amounts as long as they have soundly understood the idea.

4.5 + 1.2	add 1 and then add 0.2	**Answer:** 5.7
2.9 + 4.8	add 5 and then subtract 0.2	**Answer:** 7.7
15.6 + 8.7	add 9 and then subtract 0.3	**Answer:** 24.3

Some children at this stage may be able to extend this method into working with hundredths.

4.80 + 2.01	add 2 and then add 0.01	**Answer:** 6.81
5.75 + 1.99	add 2 and then subtract 0.01	**Answer:** 7.74

It becomes much easier if you think of the numbers as money. Try the above examples again, this time thinking in terms of pounds and pence, Euros and cents, dollars and cents, or any other decimal currency of your choice.

A COMMON CONFUSION

Something that can confuse people is a decimal calculation such as:

3.4 + 0.08

What we need to see here is that 3.4 is exactly the same as 3.40 (that is 3 **units**, 4 **tenths** and 0 **hundredths**). So we can write:

3.40 + 0.08 which, it is now much easier to see, equals 3.48

Rounding

Rounding (see also page 39) is used to **approximate** answers. However, sometimes exact answers aren't necessary. This isn't about being lazy, it's about real life. And it's about checking calculations too.

For example, 2341 tickets are sold for Accrington Stanley's midweek match and 8921 for the Saturday game. Meanwhile Plymouth Argyle sold 3019 for their midweek match and 7124 for the Saturday fixture. Which club sold the most tickets that week?

To answer that question, we don't need to do the exact sums. Rounding the numbers to get an approximate answer will be sufficient. In fact, because it makes things easier, it may point the way to a correct answer where an attempt to do a 'proper' calculation doesn't:

2341 + 8981 is approximately 2300 + 9000
and
3019 + 7124 is approximately 3000 + 7100

2300 + 9000 = 11300
and
3000 + 7100 = 10100

Without knowing the exact numbers of tickets sold, we can answer the question. Accrington sold the most tickets that week. In this example, all the numbers were rounded to the nearest 100 (see also page 40).

Apart from making some calculations easier and more manageable, rounding is very important for checking that answers are reasonable.

For example, Samir does the sum 3024 + 1207 and arrives at the answer 3231.

A quick approximation of the answer would be 3000 + 1200, giving an approximate answer of 4200. Samir can clearly see the calculated answer needs re-checking.

Negative numbers are numbers smaller than zero and are discussed in detail in Chapter 1 (see page 35).

Adding with negative numbers

The Number Line is just as useful (if not more so) when we tackle the idea of adding with negative numbers. (I *really* do recommend pupils use Number Lines as an *aide-memoire* for adding with negative numbers for a long, long time – at least until they've successfully completed their GCSEs.)

When adding we move to the *right* along the Number Line.

$$-6 \quad -5 \quad -4 \quad -3 \quad -2 \quad -1 \quad 0 \quad 1 \quad 2 \quad 3 \quad 4$$

For example:

-6 + 8
Using the Number Line above, put your finger on −6 and, *as we are adding,* jump along to the right 8 spaces.
Answer: 2

Now try:

-5 + 2
Put your finger on −5 and jump along to the *right* 2 spaces.
Answer: −3

And:

-4 + 4
Start at −4 and jump along 4 spaces to the *right.*
Answer: 0

With the help of the Number Line, it's just as easy adding negative numbers as any other numbers. With smaller negative numbers, you just need to extend the Number Line further:

$$-12\ -11\ -10\ -9\ -8\ -7\ -6\ -5\ -4\ -3\ -2\ -1\ \ 0\ \ 1\ \ 2\ \ 3\ \ 4\ \ 5\ \ 6\ \ 7\ \ 8\ \ 9$$

So,

$$-11 + 15 = \ \ 4$$
$$-4 + 12 = \ \ 8$$
$$-9 + \ \ 7 = -2$$
$$-10 + 11 = \ \ 1$$
$$-12 + 12 = \ \ 0$$
$$-12 + 21 = \ \ 9$$

At some point around the time they start secondary school, some children will be asked to think about examples such as:

8 add −5   which can also look like this:

8 + −5   or

8 + (−5)

This becomes easier when we recall that addition is commutative, that is 8 + −5 is the same as −5 + 8. Seeing the numbers in this order helps us work this out using the Number Line as before:

8 + −5   is the same as

−5 + 8   which equals

3

Alternatively we can use a money scenario. Look at this example:

5 + (−3)

Imagine the 5 as money we **own** (£5 cash in our wallet, say) and the (−3) as money we **owe** (a debt of £3). Overall we would actually have a credit of £2.

5 + (−3) = 2

Here are some more examples:

$$7 + -2 \quad \text{which equals } 5$$
$$8 + -5 \quad \text{which equals } 3$$
$$12 + -8 \quad \text{which equals } 4$$
$$0 + -2 \quad \text{which equals } -2$$

So when adding a negative number, it appears as if we are subtracting.

This is often expressed to children as:

'A plus and a minus make a minus.'

For example:

$$6 + -2 = 6 - 2 = 4$$

And this is true and helpful for some children *if* they can follow the rule.

But for this 'rule' to work the plus sign ('+') and the minus sign ('–') must be adjacent (touching or next to each other with nothing in between). This is where some children get confused and make mistakes.

- So, 8 + −5 *is* the same as 8 − 5, which equals 3.

But this 'rule' would *not* apply to a sum such as: −8 + 5
Yes, there is a plus and a minus, but they are not adjacent.

- For −8 + 5 we can use the Number Line as normal: finger on negative 8, add 5 by jumping 5 places to the *right*.
Answer: −3.

It may seem a little more tricky when children are asked to consider what happens if we add a negative number to another negative number, but the same rules apply.

For example:

$$-2 + (-3)$$

In this case we can think of -2 as being a debt of £2 and -3 as being a debt of £3. What happens if we lump or add our debts together? Well we will have a total debt of £5 (or -5).

$$-2 + -3 = -5$$

Or using the '*plus and a minus makes a minus*' rule, the above could also be rewritten as:

$$-2 - 3$$

which equals:

$$-5$$

Here is another example:

$$-3 + ? = -9$$

Keeping with our debt scenario, if we start with a debt of £3 what further debt must we add to end up with a total debt of £9? **Answer**: another debt of £6 (or -6).

$$? = -6$$

And some more examples:

$$-1 + -5 = -6$$
$$-4 + -8 = -12$$
$$-3 + -7 = -10$$
$$-9 + -3 = -12$$
$$-12 + -12 = -24$$
$$-12 + -15 = -27$$

So that is adding all summed up! I hope it has helped you have a clearer idea of what is happening in your child's classroom and also maybe reminded you of one or two things along the way. The opposite – or 'mirror image' – to addition is subtraction, and that is explored in detail in the next chapter.

3.
SUBTRACTION

So here's the second of the 'four operations of number' – **subtraction**. Also known as 'taking away', subtraction is the opposite operation to addition: it will simply 'undo' or reverse addition. As subtraction is, in effect, the 'mirror image' of addition, this chapter will often reflect and echo the previous one.

UNDERSTANDING THE BASICS
(RECEPTION AND YEARS 1 AND 2, AGES 5–7)

Children are likely to hear and use many terms that all relate to taking away: subtraction, subtract, take away, take, difference, difference between, leave, minus, counting back, fewer, less, less than, subtraction table, decrease, remove, how many are gone, how many are left, how many less than…

As with addition, your children will subtract using practical and mental methods only in the very early years. Building their confidence with numbers will be the main aim. The subtract ('–') and equals ('=') signs will only be introduced when children are familiar with the associated words and understand the operation of subtraction. This may be some time during Year 1; sooner for some, later for others. To introduce the symbols too early would be counter-productive.

Although your child might not yet have been introduced to the subtract and equals signs, I have used them throughout this chapter in order to be concise.

For very young children simply replace the symbols, in the examples below, with the words 'take away' and 'equals'.

In the classroom children will experience plenty of 'maths chat', 'mental starters', 'mental warm-ups', 'brain teasers' and 'challenges'. These are short activities and discussions designed to reinforce previously taught material and introduce new ideas.

One of the key concepts children need to be taught is that **subtraction is non-commutative.** For example: 7 – 2 is **NOT, NOT, NOT** the same as 2 – 7. (Your children won't need to learn this word, but it is an essential rule to know.)

It looks simple, and perhaps even obvious. But it may not be clear to children, especially since when we're adding up it *doesn't* matter in which order the numbers appear (see page 74). So it's really important that we make sure children see and understand that when we're taking away it *does* matter which number we start with.

Probably the best way for children to appreciate this is with real, tangible things. Start with 7 Cheerios (or raisins, Lego bricks, whatever is to hand) and ask them to take away 2. Easy, no problem, they can see they have 5 left. Repeat, this time starting with 2 Cheerios. Ask them to take away 7. Obviously they can't and will probably say it's impossible (in whatever terms a 5 or 6 year old might use).

For now, this is probably a perfectly good answer. Simply acknowledging the answer is not 5 is what we would want at this stage. If your children seem to be receptive, you may want to gently introduce the idea of negative numbers at this point. Or this can wait for another day (see also page 35).

The Number Line is a very simple and effective tool that is every bit as essential for understanding taking away as it is for adding.

When we want to subtract using the Number Line, we jump along the *spaces* to the *left*.

If children are being asked to 'do 7 take away 2', they should start by putting their finger on number 7 and then jump their finger two *spaces* to the *left*, ending up at the answer 5.

$$7 - 2 = 5$$

As with addition, it's very important to encourage children to count the 'spaces' or 'jumps' between the numbers, rather than the numbers themselves.

Here is an example:

- 'There are 10 Smarties in a new packet. If you eat 4 Smarties how many are left?'

To answer this question correctly children should put their finger on the 10 and jump back 4 *spaces*. This shows there are 6 Smarties left.

Unfortunately, some children will look at this bit of the Number Line

and give the *incorrect* answer of 7. They have counted the numbers instead of the *spaces* (or jumps) *between* the numbers and arrived at the wrong answer. This mistake is fairly common.

Number bonds

Subtraction number bonds (also known as **subtraction facts**). As for addition, all this means is how numbers 'bond' or connect to each other. For example, all the subtraction facts for the number 12 are listed below:

$$12 - 0 = 12 \qquad 12 - 7 = 5$$
$$12 - 1 = 11 \qquad 12 - 8 = 4$$
$$12 - 2 = 10 \qquad 12 - 9 = 3$$
$$12 - 3 = 9 \qquad 12 - 10 = 2$$
$$12 - 4 = 8 \qquad 12 - 11 = 1$$
$$12 - 5 = 7 \qquad 12 - 12 = 0$$
$$12 - 6 = 6$$

Likewise, for the number 19:

19 − 0 = 19	19 − 10 = 9
19 − 1 = 18	19 − 11 = 8
19 − 2 = 17	19 − 12 = 7
19 − 3 = 16	19 − 13 = 6
19 − 4 = 15	19 − 14 = 5
19 − 5 = 14	19 − 15 = 4
19 − 6 = 13	19 − 16 = 3
19 − 7 = 12	19 − 17 = 2
19 − 8 = 11	19 − 18 = 1
19 − 9 = 10	19 − 19 = 0

These bonds are very helpful to know by heart, allowing basic subtraction to happen easily. A good aim would be for children to know these number bonds for all the numbers up to 20 by around Years 3 or 4 (age 8 or 9 years old).

As mentioned earlier, most children will probably be introduced to the subtract sign ('−') in Year 1 at school and the equals sign ('=') possibly a little earlier. Children may start relating their 'maths chat' problems to subtract sums, using these symbols. This is often referred to as writing number sentences.

Subtraction number sentences use the '−' and '=' symbols and are a simple way of starting to write down, or record, mental calculations.

Imagine Jack had been doing some taking-away sums. He started with 10 marbles, gave 5 to Rafique and then gave 3 to Olivia and he worked out how many he has left. '*Good work Jack!*' says the teacher. '*Can you now record that as a number sentence?*'

Jack writes:

10 marbles − 5 marbles − 3 marbles = 2 marbles

Or he might just write:

10 − 5 − 3 = 2

It's as simple as that.

Using symbols: Another way of writing number sentences is with **symbols**. Teachers will use symbols such as ■ and ▲ to stand for numbers. For example, in increasing order of difficulty:

8 − 5 = ■	Answer: ■ = 3
19 − ▲ = 7	Answer: ▲ = 12
⬣ − 4 = 13	Answer: ⬣ = 17
⬣ − ▲ = 9	Possible answers: ⬣ = 9 ▲ = 0
	or: ⬣ = 10 ▲ = 1
	or: ⬣ = 11 ▲ = 2
	or: ⬣ = 12 ▲ = 3
	or: ⬣ = 13 ▲ = 4
	...and so on

This is another way of asking different kinds of subtraction questions and another tool for securing the **non-commutative** nature of subtraction (that is 8 − 5 is *not* the same as 5 − 8). In fact, this way of setting out a sum may be easier for some children to understand than questions in more wordy, convoluted language.

This is also a gentle introduction to **algebra**. Some adults fear algebra. Yet there really is no need: it is simply a shorthand language for mathematics. (But if you do feel the fear, never admit it to your children! Let's not plant the expectation that algebra is difficult into a child's mind.)

Nothing added, nothing taken away

Just as adding 0 (zero or nought) *always* leaves a number unchanged, so does taking away 0. It's a rather abstract idea until your children understand that taking away nothing does nothing.

For example:

- Jamal has 6 toys and he gives none away. He still has 6 toys.
  6 − 0 = 6

Sometimes looking at patterns helps children's understanding.

- Jamal has 6 toys and he gives 5 away. $6 - 5 = 1$
- Jamal has 6 toys and he gives 4 away. $6 - 4 = 2$
- Jamal has 6 toys and he gives 3 away. $6 - 3 = 3$
- Jamal has 6 toys and he gives 2 away. $6 - 2 = 4$
- Jamal has 6 toys and he gives 1 away. $6 - 1 = 5$
- Jamal has 6 toys and he gives 0 away. $6 - 0 = 6$

Partitioning

Here is a quick recap of partitioning (see also page 74), as we will need to use it as a tool in the next section. All it means is separating a number into its component parts – convenient bits that are easier to handle individually.

When subtracting two 2-digit numbers, children will be taught to partition (or separate) the second number.

In the example below, the number 17 is **partitioned** into 10 and 7.

So the sum 38 – 17 can now be rewritten as:

$$38 - 17 \rightarrow 38 - 10 - 7 \rightarrow 28 - 7 = 21$$

Similarly, when subtracting two 3-digit numbers the second number is partitioned, for example:

$$143 - 121 \rightarrow 143 - 100 - 20 - 1 \rightarrow 43 - 20 - 1 \rightarrow$$
$$23 - 1 = 22$$

The Hundred Square

We've already seen the Hundred Square:

1	2	3	4	5	6	7	8	9	10
11	12	13	14	15	16	17	18	19	20
21	22	23	24	25	26	27	28	29	30
31	32	33	34	35	36	37	38	39	40
41	42	43	44	45	46	47	48	49	50
51	52	53	54	55	56	57	58	59	60
61	62	63	64	65	66	67	68	69	70
71	72	73	74	75	76	77	78	79	80
81	82	83	84	85	86	87	88	89	90
91	92	93	94	95	96	97	98	99	100

In subtraction, children use the Hundred Square for counting back. For example, if asked to do the sum 19 − 7, your children will be taught to put their finger on the number 19 and count back (that is to the *left*) 7 places to reach the answer: 12.

A Hundred Square is great for spotting patterns, and spotting patterns is a great way to 'discover' answers. A simple but good example of this is, '*If we start with any number, what happens when we subtract 10?*'

$$34 - 10 = 24$$
$$82 - 10 = 72$$
$$67 - 10 = 57$$
$$75 - 10 = 65$$

Your children can 'discover' that while the **units** stay the same, the **tens** change. And with practice they will also be able to 'see' that to subtract 10 you just need to look up the column at the number immediately above.

I	2	3	4	5	6	7	8	9	10
11	12	13	14	15	16	17	18	19	20
21	22	23	24	25	26	27	28	29	30
31	32	33	34	35	36	37	38	39	40
41	42	43	44	45	46	47	48	49	50
51	52	53	54	55	56	57	58	59	60
61	62	63	64	65	66	67	68	69	70
71	72	73	74	75	76	77	78	79	80
81	82	83	84	85	86	87	88	89	90
91	92	93	94	95	96	97	98	99	100

When subtracting you *must* start with the first number. For example with 78 – 13, start by putting a finger on 78. Initially children will probably just count back 13 squares and arrive at the answer 65. But by the age of 7 or so, your children may be partitioning numbers first.

For example: 78 – 53. They put their finger on 78. Then they partition the other number into **tens** and **units** (53 → 50 + 3). To subtract the tens (50) they move their finger up the columns. They move 5 columns in this case because there are 5 tens (50).

So their finger is now on 28 (because 78 – 50 → 28). Then, to subtract the units, they move their finger along the row to the *left*, counting back 3 places to arrive at the answer: 25.

I	2	3	4	5	6	7	8	9	10
11	12	13	14	15	16	17	18	19	20
21	22	23	24	25	26	27	28	29	30
31	32	33	34	35	36	37	38	39	40
41	42	43	44	45	46	47	48	49	50
51	52	53	54	55	56	57	58	59	60
61	62	63	64	65	66	67	68	69.	70
71	72	73	74	75	76	77	78	79	80
81	82	83	84	85	86	87	88	89	90
91	92	93	94	95	96	97	98	99	100

In summary: to subtract the **tens** travel up the columns, to subtract the **units** travel to the *left* along the rows.

When using the Hundred Square, your children may 'see' or discover easier ways to do things. For example, an easy way to subtract 9 would be to subtract 10 and then add on 1. Similarly, to subtract 11 it is easy to subtract 10, then another 1. These are examples of **adjusting**.

Adjusting

This technique is all about making things simpler. Some sums are simply easier to do in our heads than others. So, we start by doing what we find easier and then '**adjust**' afterwards – to make the sum 'work'. A couple more examples will show you what I mean.

Subtracting 20 from a number is usually quite straightforward. So, if we needed to subtract 19 from any number we could start by subtracting 20 and then **adjust** afterwards by adding on 1. For example:

37 – 19
Do 37 – 20, which is 17
and then **adjust** by adding on 1
(because we've taken away 1 too many)
17 + 1 gives us
18

To subtract 21 from any number we could start by subtracting 20 and then **adjust** by subtracting another 1. For example:

68 – 21
Do 68 – 20 = 48
and then adjust by subtracting another 1
48 – 1 gives us
47

Obviously this can be extended to bigger numbers and greater adjustments. For example:

378 − 102
Do 378 − 100 and then adjust by subtracting another 2
278 − 2 = 276

524 − 97
Do 524 − 100 and adjust by adding 3 back on
424 + 3 = 427

As mentioned earlier, subtraction is the **opposite operation** to addition. In other words, subtraction and addition **reverse** or **undo** each other. (Technically speaking, subtraction and addition are the **inverse** of each other.)

As   15 − 3 = 12
then   12 + 3 *must equal* 15

So, if you've subtracted 3 by mistake, to get back to where you started, simply add 3. This may sound very straightforward, but it often needs pointing out to children and it is a *very helpful* thing to know, especially when they go on to deal with bigger numbers and more complex sums (and when using a calculator, for example).

If children realise they have made a mistake midway through a more complex sum, it's a lot more efficient and rewarding to be able to go back just one stage than having to start the whole sum again. It's also a good way to check a sum, and sometimes even to calculate it in the first place, as we'll see below.

In the early days, children are often more confident with addition than subtraction. Being able to see the connection between the two can help them become more competent with subtraction.

Counting up or **counting on** (adding) can be an easier option than counting down (subtracting), especially when the numbers are close together. This is also known as **complementary addition** – making addition and subtraction work hand in hand.

For example:

83 – 79

To find the **difference** between these two numbers and therefore the answer to the sum, picture the Number Line:

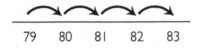

79   80   81   82   83

and **count up** the number of jumps from 79 until you reach 83. Easy. Answer: 4

Likewise with **602 – 596**

596   597   598   599   600   601   602

We can start at 596 and count up until we reach 602, thereby finding the **difference** between the two numbers.

Answer: 6

> Difference is a term your children *will* be expected to know. It simply means to find out how far apart two numbers are. It is often used in verbal questions.

For example:

- 'Find the difference between 96 and 84.'
- 'What's the difference in ages between you and your oldest sister?'
- 'What's the difference in ages between you and your youngest brother?'
- 'What's the difference between 19 and 39?'

As you can see, with the term **'difference'** it does *not* matter in which order the numbers are presented. It is simply a means of finding out how far apart the numbers are.

Related subtraction facts

Using **related subtraction facts** simply means using things we know to help us solve things we don't know.

For example, to calculate 130 – 50, children may use the fact that they *know* 13 – 5 = 8, along with their knowledge of place value, to derive that:

130 – 50 = 80

Subtracting 100: Just as with taking away 10 from 2-digit numbers at an earlier stage, the process of subtracting 100 from any 3-digit number is one that is certainly worth practising again and again. Initially it is probably best to avoid using thousands. For example:

683 – 100 = 583
837 – 100 = 737
219 – 100 = 119
750 – 100 = 650   ...and so on

Later on this could be extended to:

1250 – 100 = 1150
2168 – 100 = 2068
1040 – 100 = 940
3052 – 100 = 2952   ...and so on

We saw earlier that subtraction is **non-commutative**. If children don't understand why this is true – and the earlier they do the better – we end up with wrong answers and lots of frustration.

Because subtraction is non-commutative, it is even more important to ensure children really understand the question.

Understanding the question is a skill that takes practice. Even for children who have grasped and understood the important principle that subtraction is non-commutative, questions such as '*Take 3 from 8*' or '*Subtract 4 from 11*' can cause problems.

With questions like the examples above, children often 'see' or 'hear' what they think the sum is 'saying', and not what is actually being asked.

Our brains have awesome powers of perception, but they constantly take shortcuts by picking up on part of the picture, fitting it to something familiar and leaping to conclusions. Children are not alone in this; adults do it all the time. Strangers in a crowd remind us of familiar faces, two facial profiles become a candlestick, and French Connection UK makes a fortune out of its cheeky abbreviation. We see what we *think* we ought to see.

So, when children are asked a question like '*Take 3 from 8*' they will often pick up on the cue that it is a '*Take*' sum and try to fit the order of the numbers to the familiar pattern, 3 first then 8. If they then attempt the incorrect sum of 3 – 8, they will either come up with the wrong answer or give up altogether.

It's an easy mistake and one that even children with a confident grasp of numbers can make. It happens more often when questions are written down than when they are asked verbally, but either way it's important that we are aware of why children are getting this kind of question confused so we know how to help.

The word '*from*' is the cue we need to teach. *From 8* is where children need to start and put their finger on the Number Line. So, finger on 8, *take 3* (that is 3 jumps to the *left*).

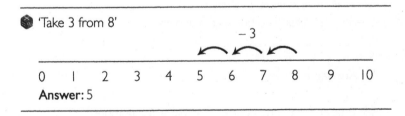

🎲 'Take 3 from 8'

– 3

| 0 | 1 | 2 | 3 | 4 | 5 | 6 | 7 | 8 | 9 | 10 |

Answer: 5

Similar questions that can cause problems – for children of all ages – are:
- 'What number must I take from 18 to leave 7?'

- 'How many more is 11 than 5?'
- '8 taken from a number is 13. What is the number?'
- '6 added to a number is 16. What is the number?'
- 'Find pairs of numbers with a difference of 9.'
- 'How many less than 38 is 14?'
- '30 taken from a number is 45. What is the number?'

So, if it's the language and not the sum that's the problem, we need to teach the whole package – unpack the words and the rest is easy.

UNPACKING THE LANGUAGE OF SUBTRACTION

Let's try unravelling some of the examples of tricky language using the Number Line.

'What number must I take from 18 to leave 7?'
In other words start at 18, and finish at 7.

| 7 | 8 | 9 | 10 | 11 | 12 | 13 | 14 | 15 | 16 | 17 | 18 |

Answer: 11

'How many more is 11 than 5?'
Find the difference between 5 and 11.

| 5 | 6 | 7 | 8 | 9 | 10 | 11 |

Answer: 6

'8 taken from a number is 13. What is the number?'
Start at 13 and *add back on* **8**

| 13 | 14 | 15 | 16 | 17 | 18 | 19 | 20 | 21 |

Answer: 21

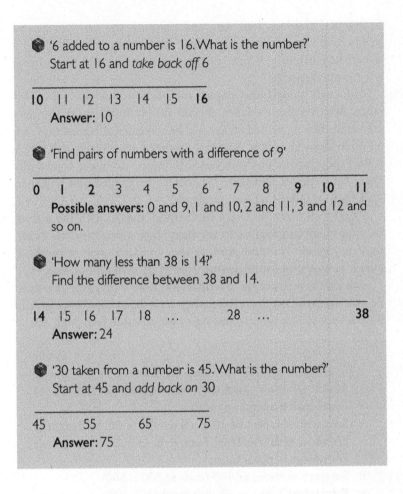

'6 added to a number is 16. What is the number?'
Start at 16 and *take back off* 6

10 11 12 13 14 15 **16**
Answer: 10

'Find pairs of numbers with a difference of 9'

0   1   2   3   4   5   6   7   8   9   10   11
Possible answers: 0 and 9, 1 and 10, 2 and 11, 3 and 12 and so on.

'How many less than 38 is 14?'
Find the difference between 38 and 14.

14 15 16 17 18 ...   28 ...   **38**
Answer: 24

'30 taken from a number is 45. What is the number?'
Start at 45 and *add back on* 30

45   55   65   75
Answer: 75

MOVING FORWARD
(YEARS 3 AND 4, AGES 7–9)

We've covered the basics in subtraction in the previous section 'Understanding the Basics', and these continue to be used as your children progress. At this point I think it's worth acknowledging that just because children have been *shown* the basics this does not mean they fully *understand* the basics. Being confident and competent with all of the above will take time and practice – and a good dollop of patience from all concerned!

Even if children have breezed through the early years and grasped all the concepts thrown at them, they will still spend a significant amount of time practising, securing and fine-tuning their mental maths.

Informal **'pencil-and-paper'** methods are introduced at this stage. Writing things down should simply build on the understanding they've gained so far and be a natural progression from all their mental maths. So, early written recordings will often reflect their mental methods but with bigger numbers. Eventually these early 'pencil-and-paper' methods are refined and will then lead on to standard written methods.

The written methods are built up in stages, with the aim of helping children progress to the next stage when they feel confident to do so. The end point is the standard column method, which you may be familiar with from your own school days.

For children to use these written methods with ease, they must be able to build on everything they have been taught so far and have a very secure understanding of **place value** (see page 30).

In summary: to subtract with ease children need to be able to:
- Recall all their **addition** and **subtraction facts (number bonds)** up to 20.
- Subtract multiples of 10, such as 130 − 80, by using the **related subtraction fact** that 13 − 8 = 5 and their knowledge of place value.
- Subtract multiples of 100 (such as 800 − 500).
- **Partition** 2-digit and 3-digit numbers in *different* ways (for example partition 84 into 80 + 4 or into 70 + 14).

The stages...

Stage 1: The Empty Number Line

The Number Line is still used at this stage, but it can become a little more abstract. It's sometimes referred to as the **Empty Number Line**. Numbers need not be written sequentially, and just a sketch of a line with a few jottings may be all that is recorded.

When we are using an Empty Number Line to subtract, we need to start at the far right of the line and then move or *jump* to the *left*. The 'scale' of the line is now irrelevant; it is simply used as a prop to aid calculation.

For example, for the sum 174 – 63, it might look something like this (remember we're jumping to the *left* along the line for subtraction):

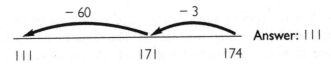

Answer: 111

Another example for 97 – 48 might look like this:

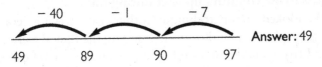

Answer: 49

And another for 238 – 57 might look like this:

Answer: 181

(Pupils at secondary school will still use Number Lines, similar to these, to help them perform and check their calculations.)

Here are some more examples you might like to try with your children: 15 – 8

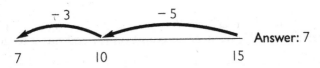

Answer: 7

84 – 27, which could look like this:

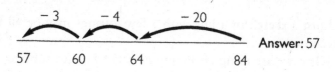

Or like this:

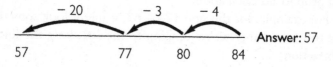

The Empty Number Line simply helps keep track of the steps involved in the mental calculation. The steps in the examples above all show counting back but subtraction can also be done by counting up.

An alternative way of using the Empty Number line at this stage is to use **counting up** *or* **counting on**.

We looked at this technique, otherwise known as **complementary addition**, in the 'Understanding the Basics' section (see page 116). Children at a slightly more advanced level can use it with bigger numbers, and will only need an Empty (or 'Abstract') Number Line.

For example:

$$306 - 297$$

All we do is imagine 297 on a Number Line and think about how many we need to count upwards (or to the *right*, I should say) to get to 306.

$$+ 9$$

297   306

So starting at 297, we need to count on (or add on) 9 to get to 306.

$$306 - 297 = 9$$

Here is another example:

84 −27

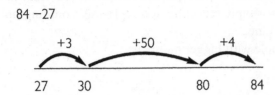

We start with 27 on the Number Line and count on in steps until we reach 84. What *we have counted up* is the answer to the sum.

Answer: 57

Children might also start recording their work like this:

$$
\begin{array}{r}
84 \\
-27 \\
\hline
\end{array}
$$

+	3	→ 30	**count up** 3 to make 30
+	50	→ 80	**count up** 50 to make 80
+	4	→ 84	**count up** 4 to make 84

Answer: 57   the total of what has been 'counted up' and therefore the answer

Or with a reduced number of steps:

$$
\begin{array}{r}
84 \\
-27 \\
\hline
\end{array}
$$

+	3	→ 30
+	54	→ 84

Answer: 57

Because this style of recording may look unfamiliar to us, it may appear confusing. A typical reaction is for parents to think their children are making things more complicated than they need. But children usually find this an easy method to master. This informal written method is simply reflecting and replacing 'counting up' using a Number Line.

Another example with bigger numbers:

	526		or		526	
−	287			−	287	
	———				———	
+	3	→ 290		+	13	→ 300
+	10	→ 300		+	226	→ 526
+	200	→ 500			———	
+	26	→ 526			239	
	———					
	239					

COUNTING UP WITH DECIMAL NUMBERS

This example simply shows that the principles set out above apply just as well to decimal numbers as to whole numbers. Again, the purpose is to find the difference between the two numbers by starting with the smaller number and 'counting up' in steps until we reach the larger number (see also page 117).

	32.4		
−	26.7		**start** with 26.7
	———		
+	0.3	→ 27.0	**count up** 0.3 to make 27.0
+	5.0	→ 32.0	**count up** 5 to make 32.0
+	0.4	→ 32.4	**count up** 0.4 to make 32.4
	———		
Answer:	5.7		the total we have **counted up** and therefore the difference between 26.7 and 32.4

This method of 'counting up' is just as effective when subtracting with decimal numbers. Decimal numbers are not whole numbers. Decimal numbers are made up of a whole number and a fractional part – separated by a decimal point. Children are unlikely to be subtracting with decimal numbers at this stage but an example is shown in the box above – for future reference.

Stage 2: Partitioning

Partitioning can be used to record the steps in a mental calculation and is a way of linking mental calculations and written calculations. In the example below, the number 34 is partitioned (or separated) into 30 and 4. So the sum 83 – **34** can now be rewritten as:

$$83 - 34 \rightarrow 83 - 30 - 4$$

Now we can subtract each bit of the 34 separately.

$$83 - 30$$

which equals 53. We can either 'keep' this bit of the answer in our head – or jot it down – and then we do:

$$53 - 4 = 49$$

Other examples (showing possible ways to partition):

$$45 - 6 \rightarrow 45 - 5 - 1 \rightarrow 40 - 1 = 39$$
$$65 - 17 \rightarrow 65 - 15 - 2 \rightarrow 50 - 2 = 48$$
$$350 - 160 \rightarrow 350 - 150 - 10 \rightarrow 200 - 10 = 190$$

Both numbers may then be partitioned and written under one another.
    For example: 87 – 23

$$80 + 7$$
$$- \quad 20 + 3 \quad \rightarrow$$

$$80 + 7$$
$$- \quad 20 + 3$$

$$60 + 4 \quad \textbf{Answer: 64}$$

And this is the link to the next stage.

Stage 3: Expanded layout leading to the column method

The column method is introduced at this stage but will probably not be fully mastered until Years 5 and 6. For each example below the **column method** is shown in bold next to the **expanded layout**.

Decomposition is another technical-sounding word but the process it describes, splitting up a number into its units, tens and hundreds, is very similar to partitioning. Decomposition actually means 'breaking down' so...

526	can be decomposed into	$500 + 20 + 6$
32	can be decomposed into	$30 + 2$
7825	can be decomposed into	$7000 + 800 + 20 + 5$
9040	can be decomposed into	$9000 + 0 + 40 + 0$

Decomposition is used to do subtraction sums and the word 'decomposition' has now become synonymous with the modern method of doing written subtractions.

DECOMPOSING DECOMPOSITION

Adults may find using decomposition to subtract takes a bit of getting used to. It's not how we did it and so it can sometimes seem awkward and clumsy. You may itch to show your children 'your' way of doing it. Please try and refrain from this because you really won't be helping.

This really is an example of something they just don't do the same way in schools any more. So for better or worse, decompose we must.

Fortunately your children do not need to use the word, just the method it describes. And the method is a sound one that does make sense – unlike the familiar old 'borrow-and-carry' method that we probably used. Those terms were quick and easy to use if we were fluent with the method, but utter nonsense if we were not.

The next section will show you how children are introduced to the idea of using decomposition to subtract.

The expanded layout is illustrated first, leading to the more formal standard column (or compact) method (shown in **bold** throughout for reference). Pupils will be introduced to the expanded layout method in Years 3 and 4, refining and becoming more efficient with their workings until, in Year 6, they will probably just use the standard or column ('**bold**') method.

So how does it work? The numbers are broken down – or decomposed – into their **hundreds, tens** and **units** and are laid out as shown below.

Units are then subtracted from **units, tens** from **tens, hundreds** from hundreds and so on.

If at any point there are not enough **units** (or **tens** or **hundreds** or …) on the top line, then further decomposition must take place.

With the 'decomposition' method (unlike the 'borrow-and-carry' method) ALL the action takes place on the top line of the sum.

In the example below 482 is decomposed into 400 and 80 and 2. Likewise 261 is decomposed into 200 and 60 and 1.

The sum is then written in an expanded layout:

$$482 - 261 \rightarrow \begin{array}{r} 400 + 80 + 2 \\ - \quad 200 + 60 + 1 \\ \hline \end{array}$$

Now is the time to encourage children to always start subtracting from the '**units**' or '**ones**', followed by the '**tens**' and then the '**hundreds**'.

Reading the numbers out loud helps, so it would be '*2 take away 1, then 80 take away 60* (not 8 take away 6) *and then 400 take away 200*' (not 4 take away 2)'.

$$482 - 261 \rightarrow \quad 400 + 80 + 2$$
$$- \quad 200 + 60 + 1 \quad \text{leading to the} \qquad 482$$
$$\overline{} \quad \text{column method} \qquad - 261$$
$$200 + 20 + 1 \qquad\qquad\qquad 221$$

For these methods to work, your children must know that **units** line up under **units**, **tens** under **tens** and so on – just like we always did at school. Sometimes this is referred to as a 'vertical layout'.

Your children will probably start with sums involving only **tens** and **units**.

For example:

$$89 - 54 \rightarrow \quad 80 + 9 \quad \text{mirrors} \qquad 89$$
$$- \quad 50 + 4 \quad \text{column} \qquad - 54$$
$$\overline{} \quad \text{method}$$
$$30 + 5 \qquad\qquad 35$$

So the children say to themselves '9 *take away 4 is 5, and 80 take away 50 is 30*'. **Answer**: 35

So far so good.

Now for one a little more challenging:

$$73 - 27 \rightarrow \quad 70 + 3$$
$$- \quad 20 + 7$$
$$\overline{}$$

Once again the children start with the **units** and say to themselves '3 *take away 7*' and find that this won't 'work'. What happens now?

Well, we need to reorganise or regroup the numbers, so that it *can* 'work'.

Basically, we need more units on the top line!

The children look to the **tens** column and know they must decompose again. They break down the 70 into '60 + 10'. The 70 is replaced with 60 and the 10 then slides across to join the original units. Now, instead of 3 **units** we have 13 **units** (10 + 3).

$$
\begin{array}{rr}
70 + 3 & \rightarrow \\
- \; 20 + 7 & \\
\hline
\end{array}
\qquad
\begin{array}{r}
60 + 13 \\
- \quad 20 + \; 7 \\
\hline
40 + \; 6 \\
\end{array}
$$

The children now say, '*13 take away 7 is 6, and 60 take away 20 is 40*'. **Answer:** 46

As children grow in confidence, they may choose to record their 'workings' in a slightly more shorthand manner. The 70 and 3 are crossed out and replaced with the 60 and 13 respectively, as shown below:

$$
73 - 27 \quad \rightarrow \quad
\begin{array}{r}
^{60} \quad ^{13} \\
\cancel{70} + \cancel{3} \\
- \quad 20 + 7 \\
\hline
40 + 6 \\
\end{array}
\begin{array}{l}
\text{this now} \\
\text{mirrors the} \\
\text{column method} \\
\end{array}
\qquad
\begin{array}{r}
^{6\,13} \\
\cancel{73} \\
- \, 27 \\
\hline
\mathbf{46} \\
\end{array}
$$

This may not seem as succinct as 'our' old method, but it does make more sense. It relies on logic and basic mathematical understanding. Children are taught from a very early age how to break down or partition numbers in different ways (in other words they *understand* that 73 is the same as 60 and 13). So, instead of borrowing a mystery '1' from somewhere and then carrying it back somewhere else (!), they know they are simply moving '10' from the **tens** column into the **units** column.

Now for sums involving **hundreds, tens** and **units**...

An easy one to start with:

$$
793 - 582 \quad \rightarrow \quad
\begin{array}{r}
700 + 90 + 3 \\
- \; 500 + 80 + 2 \\
\hline
200 + 10 + 1 \\
\end{array}
\quad \text{leading to} \quad
\begin{array}{r}
793 \\
- \, 582 \\
\hline
211 \\
\end{array}
$$

No problem! (As no further decomposition or adjustment is needed.)

Now for a more challenging example: 384 – 156

$$384 - 156 \rightarrow \begin{array}{r} 300 + 80 + 4 \\ - \quad 100 + 50 + 6 \\ \hline \end{array}$$

Children will attempt '4 subtract 6' and realise they can't 'do' it. There just aren't enough **units** on the top line. So they will break down (or decompose) 80 into '70 + 10'. Now the 80 will be replaced by 70 and the 10 will slide across into the **units** column and be added to the original 4, making 14 **units**.

Now they can do '14 subtract 6', giving 8. The rest of the sum is easy: 70 subtract 50 is 20; 300 subtract 100 is 200. **Answer: 228**

So in summary:

$$384 - 156 \rightarrow \begin{array}{r} 300 + 80 + 4 \\ - \quad 100 + 50 + 6 \\ \hline \end{array} \rightarrow \begin{array}{r} 300 + 70 + 14 \\ - \quad 100 + 50 + 6 \\ \hline 200 + 20 + 8 \end{array}$$

Again, as children grow in confidence, they may prefer the short-hand version. The 80 and 4 are crossed out and replaced with the 70 and 14 respectively, as shown below:

$$384 - 156 \rightarrow \begin{array}{r} \overset{70 \quad 14}{300 + \cancel{80} + 4} \\ - \quad 100 + 50 + 6 \\ \hline 200 + 20 + 8 \end{array}$$

This is now just one stage away from the standard column method.

Having mastered the expanded layout method of doing subtraction, your children will have a good understanding of 'how and why' the sum works. At this point (maybe in Year 5 but earlier for some and later for others), your children will be 'ready' for the standard column method:

$$7\ 14$$

384 – 156 → **384**   Once again the 80 and 4 are
                    – **156**   crossed out and replaced with the
                    ——   70 and 14 respectively.
                    228

Two more examples for you to peruse:

1)   835 – 274

$$800 + 30 + 5$$
$$-\ 200 + 70 + 4$$
————————

Adjustment needed from **hundreds**.
   Children may show their 'working' like this:

$$700 + 130 + 5$$ as 800 + 30 is the same as 700 + 130
$$-\ 200 +\ 70 + 4$$
————————
$$500 + 60 + 1$$

and / or

$$700\qquad 130$$
$$800 + 30 + 5$$
– $$200 + 70 + 4$$
————————
$$500 + 60 + 1$$

This will eventually lead on to the standard column:

$$7\ 13$$

835 – 274 → **835**   The 800 and 30 are crossed out and
                    –274   replaced with 700 and 130 respectively.
                    ——
                    561

2)   951 − 365

$$900 + 50 + 1$$
$$-\ \ 300 + 60 + 5$$
$$\overline{}$$

Adjustment needed from tens and hundreds.
Children may show their 'working' like this:

$900 + 40 + 11$		$800 + 140 + 11$
$-\ \ 300 + 60 + 5$ and		$-\ \ 300 + 60 + 5$
——————— then		———————
6		$500 + 80 + 6$

Initially the 50 + 1 is partitioned into 40 + 11 and we can start to tackle the sum: 11 − 5 = 6. But then we look at the 40 − 60 and can't 'do' it, so we need to adjust again.

The 900 + 40 is *then* partitioned into 800 and 140 .

Now the sum can be completed (working right to left – the opposite to reading):

11 − 5 = 6; 140 − 60 = 80; 800 − 300 = 500,
giving the answer 586.

Children may also show their working like this:

800   140
~~40~~   11
~~900~~ + ~~50~~ + ~~1~~   the 50 and the 1 are crossed out and
300 + 60 + 5   replaced with 40 and 11; *then* the
———————   900 and the 40 are crossed out and
500 + 80 + 6   replaced with 800 and 140

This will eventually lead on to the standard column:

```
    8 14
      4 11
    9̶5̶1̶
  - 365
   ───
    586
```

What happens if we want to decompose but there are no **tens** or no **hundreds** from which to do so? (When 0 is a place holder.)

Look at this example: 602 − 385

We can set it out as with all the other examples shown above:

$$
\begin{array}{r}
600 + \ 0 + 2 \\
- \quad 300 + 80 + 5 \\
\hline
\end{array}
$$

In an attempt to do '2 *take away* 5' we soon realise we need to decompose again, so look to the **tens** column. But there are no **tens**.

In this example the 0 is acting as a place holder for the **tens** so decomposition has to be done in two stages: 600 is first partitioned into '500 + 100'; and *then* the 100 is partitioned into '90 + 10'. The 10 can now slide across to join the original 2 units giving 12 units.

$$
\begin{array}{r}
600 + \ 0 + 2 \\
- \quad 300 + 80 + 5 \\
\hline
\end{array}
\quad \rightarrow \quad
\begin{array}{r}
500 + 100 + 2 \\
300 + \ 80 + 5 \\
\hline
\end{array}
\quad \rightarrow \quad
\begin{array}{r}
500 + 90 + 12 \\
300 + 80 + \ 5 \\
\hline
200 + 10 + \ 7
\end{array}
$$

Eventually children may show the sum like this:

```
                90   12
         500    1̶0̶0̶                        5 9 12
        6̶0̶0̶ +  0̶ + 2̶                       6̶0̶2̶
    -   300 + 80 + 5     leading to      - 385
        ───────────                        ───
        200 + 10 + 7                        217
```

Is it reasonable?

It's a really good idea to get your children into the habit of checking whether their answer is reasonable. As the sums become more complex, like those above, checking becomes even more important. They should ask themselves '*Is my answer sensible?*'

As we touched on earlier, remember that subtraction will 'undo' addition and vice versa. With a little guidance, your children can learn to use this as an important way to **check their answers**.

- From the sum above we have: $602 - 385 = 217$
- If '217' is the correct answer, then: $217 + 385$ should equal 602

So to check:

$$\begin{array}{r} 217 \\ + \ 385 \\ \hline 602 \\ \hline \scriptstyle 1\ 1 \end{array}$$

Yes!

This method also works well when doing mental maths.

For example, if asked: $78 - 13$ you may think the answer is 65.

To check that it *is* the correct answer, you can add the 13 back on to 65 to check that it takes you back to 78. (Yes: $65 + 13 = 78$.)

TO THE TOP
(YEARS 5 AND 6 PLUS, AGES 9–12 PLUS)

As your children continue on their mathematical journey, all the rules and methods for subtraction that we've covered in the sections above are still in daily use. Expect bigger (and smaller) numbers and more complex questions, but the same principles are still at work. In this section, we familiarise ourselves with some more terms and methods, and make the move into decimals and negative (minus) numbers.

Compensation is a technical term for a simple idea: taking away a number that is easier to subtract but is too big, then adding back on (**compensating**) afterwards.

For example, with 652 − 84 you might start by subtracting 100 instead of 84.

And 652 − 100 = 552.

100 is an easier number to subtract but it does mean we have subtracted too much.

Now we know the difference between 84 and 100 is 16, so at this stage we would have subtracted 16 too many.

We **compensate** for this on the next line by adding 16 back on. 652 − 84 is written as:

```
    652
 −   84
    ───
    552     Start by subtracting an 'easier' number such as 100:
 +   16     then compensate by adding 16 back on
    ───     (because the difference between 100 and 84 is 16)
    568
```

Another example: 321 − 192 is written as:

```
    321
 −  192
    ───
    121
 +    8
    ───
    129
```

We started by subtracting 200 because this is an easier number to deal with than 192. This gave us: 321 − 200 = 121. But because 200 is a bigger number than 192 we have taken away too many (8 too many to be exact), so we **compensate** by *adding* 8 back on.

The column (or compact standard) written method

By about age 9, most children will have been taught to subtract units, tens and hundreds in columns using the **decomposition** method (see page 128 for a detailed explanation). As they get older and grow more confident in applying it, the numbers used in the sums will usually get bigger and more stages of decomposition may be involved. Here are some examples:

$586 - 58$

$$
\begin{array}{r}
{}^{7\,16}5\cancel{86} \\
-58 \\
\hline
528
\end{array}
$$

86 is decomposed into 70 + 16

$425 - 34$

$$
\begin{array}{r}
{}^{3\,12}4\cancel{25} \\
-34 \\
\hline
391
\end{array}
$$

400 and 20 is decomposed into 300 + 120

$362 - 83$

$$
\begin{array}{r}
{}^{2\,15}_{\ \ \cancel{5}\,12}3\cancel{62} \\
-83 \\
\hline
279
\end{array}
$$

60 and 2 is decomposed into 50 + 12
then 300 and 50 into 200 + 150

$842 - 349$

$$
\begin{array}{r}
{}^{7\,13}_{\ \ \cancel{3}\,12}8\cancel{42} \\
-349 \\
\hline
493
\end{array}
$$

$$\begin{array}{r}
{\scriptstyle 1\ 14} \\
{\scriptstyle 4\ 15} \\
\cancel{2557} \\
-\ 784 \\
\hline
1773
\end{array}$$

2557 − 784

$$\begin{array}{r}
{\scriptstyle 7\ 15} \\
{\scriptstyle 6\ 15} \\
\cancel{7865} \\
-\ 4287 \\
\hline
3578
\end{array}$$

7865 − 4287

Children must realise that it is vital to arrange **units** under **units**, **tens** under **tens** and so on in order to get the right result.

Subtracting with decimals

A good way to introduce the idea of subtracting decimals may be to start with money.

For example: 95p − 23p can be also be shown as £0.95 − £0.23

Children will first be expected to take away in their head (mental methods) before attempting the standard written methods.

Something that can throw people is a decimal calculation such as:

3.4 − 0.08

What we need to see here is that 3.4 is exactly the same as 3.40 (that is, **3 units, 4 tenths** and **0 hundredths**). So we can write

3.40 − 0.08

This is now easier to tackle.

Answer: 3.32

Think of it as money. *£3.40 take away 8p. Answer: £3.32.*

Children would probably be expected to do a sum such as this in their heads. Using everyday contexts and money certainly helps a lot of children 'know' where to start.

You would be amazed (I still continue to be so) at the difference in a child's apparent 'ability' when *seeing* a question written down compared with *hearing* it in context.

Many secondary-age children would completely refuse to attempt such a sum as: 5.65 – 1.9. Yet if you *say* to them: '*What is £5.65 take away £1.90?*' they can often give you the answer in seconds: £3.75.

This is not cheating or doing it for them – it is simply helping children realise what they can do. (And also reaffirming that 1.9 is exactly the same as 1.90 or for that matter the same as 1.900 or 1.9000 or 1.9000…).

The **standard column** (or **compact**) written method for subtracting decimals is the same as we've seen above. Children just need to know and remember a few extra things:

- The decimal points must *always* line up underneath each other.
- Exactly as the **hundreds**, **tens** and **units** need to line up in straight columns, so do the **tenths** and **hundredths** (and **thousandths** and so on).

So 0.95 – 0.23 would look like this:

$$
\begin{array}{r}
0.95 \\
-\ 0.23 \\
\hline
0.72 \\
\hline
\end{array}
$$

It is particularly important to remember this when subtracting sums with different numbers of decimal places (which children in

Year 6 and beyond will be expected to do). For example:

$$342.8 - 18.25 \quad \text{or} \quad 26.34 - 9.8$$

Should look like:

$$
\begin{array}{r}
342.8 \\
-\ 18.25 \\
\hline
\end{array}
\qquad
\begin{array}{r}
26.34 \\
-\ 9.8 \\
\hline
\end{array}
$$

Children need to be taught that a number such as 342.8 is exactly the same as 342.80 and 342.800 and 342.8000 and so on. Likewise 9.8 is the same as 9.80 is the same as 9.800 is the same as 9.8000 is the same as… you get the idea. The extra zeros may be needed to help ensure the correct layout when two numbers have different numbers of decimal places (see also page 100).

So the above sums can now be written as:

$$
\begin{array}{r}
3\ 12\ 7\ 10 \\
3\cancel{42.80} \\
18.25 \\
\hline
324.55 \\
\hline
\end{array}
\qquad
\begin{array}{r}
1\ 15 \\
\cancel{5}\,13 \\
\cancel{26.34} \\
9.80 \\
\hline
16.54 \\
\hline
\end{array}
$$

Adjusting

The same method of simplifying a sum as explained in 'Understanding the Basics' (see page 115) works just the same with bigger numbers and with decimal numbers. You simply subtract a similar, but more convenient number and then adjust the result up or down. For example:

Subtract 9, 19, 29, 39… or 11, 21, 31, 41… from any number by subtracting 10, 20, 30, 40… respectively and adjusting by 1. This method makes it very much easier to do these sums *in your head*.

56 –	19	subtract 20 and then add back on 1	Answer: 37
62 –	31	subtract 30 and then subtract another 1	Answer: 31
356 –	39	subtract 40 and then add back on 1	Answer: 317
673 –	81	subtract 80 and then subtract another 1	Answer: 592
1256 –	99	subtract 100 and then add back on 1	Answer: 1157
4583 –	201	subtract 200 and then subtract another 1	Answer: 4382
3256 –	1999	subtract 2000 and then add back on 1	Answer: 1257
8739 –	4001	subtract 4000 and then subtract 1	Answer: 4738

Once the idea of adjusting has been grasped, children will be happy to adjust by more than 1.

82 –	23	subtract 20 and then subtract another 3	Answer: 59
67 –	18	subtract 20 and then add back on 2	Answer: 49
374 –	96	subtract 100 and then add back on 4	Answer: 278
2025 –	1995	subtract 2000 and then add back on 5	Answer: 30
6573 –	2006	subtract 2000 and then subtract 6	Answer: 4567

Similarly for decimal numbers, subtract 0.9, 1.9, 2.9, 3.9... or 1.1, 2.1, 3.1, 4.1... from any number by subtracting 1, 2, 3, 4... respectively and adjusting by 0.1

2.6 – 0.9	subtract 1 and then add back on 0.1	Answer: 1.7
8.7 – 4.9	subtract 5 and then add back on 0.1	Answer: 3.8
12.3 – 7.1	subtract 7 and then subtract another 0.1	Answer: 5.2
32.9 – 2.1	subtract 2 and then subtract another 0.1	Answer: 30.8
78.6 – 3.1	subtract 3 and then subtract another 0.1	Answer: 75.5

Again, children can learn to adjust by greater amounts as long as they understand the idea soundly.

4.5 – 1.2	subtract 1 and then subtract another 0.2	Answer: 3.3
32.9 – 4.8	subtract 5 and then add back on 0.2	Answer: 28.1
15.6 – 8.7	subtract 9 and then add back on 0.3	Answer: 6.9

Some children at this stage may be able to extend this method into working with hundredths.

4.80 − 2.01   subtract 2 and then subtract 0.01   Answer: 2.79
5.75 − 1.99   subtract 2 and then add back on 0.01   Answer: 3.76

It becomes much easier if you think of the numbers as money (something that even young children relate to quite easily). Try the above examples again, this time thinking in terms of pounds and pence.

Rounding and approximation

Rounding (see also 39) is used to **approximate** answers. As we saw in the previous chapter, sometimes exact answers aren't necessary. Approximate answers are often sufficient, especially in real-life situations. And being able to approximate answers is a good way of checking calculations and ensuring exact answers are sensible.

For example, Annie has two savings accounts, and two big fat credit card bills to pay off. The first savings account has £9341 in it and the second has £8019. One credit card balance is £7124 and the other is £6981. If she used each savings account to clear each card respectively, which account would have most left in it after clearing her debts?

To answer that question, she could do the exact sums, but it isn't really necessary. Rounding the numbers to get an approximate answer will be sufficient. In fact, because it makes things easier, it may point the way to a correct answer where an attempt to do a 'proper' calculation does not:

     9341 − 7124 is approximately 9300 − 7100
and
     8019 − 6981 is approximately 8000 − 7000

     9300 − 7100 = 2200
and
     8000 − 7000 = 1000

Without knowing the exact pounds and pence, we can answer the question. In this example, all the numbers were rounded to the nearest 100.

Apart from making some calculations easier and more manageable, rounding is very important for checking that answers are reasonable.

For example, Annie does the sum 9341 – 7124 and arrives at the answer 9217. She's missed the digit '7' when tapping the sum into the calculator.

As we saw above, a quick approximation of the answer would be 9300 – 7100, giving an approximate answer of 2200. Knowing this, she can clearly see the calculated answer needs re-checking, and tries the sum again.

Answer: 2217

Numbers smaller than 0 are called negative numbers (see also page 35).

Subtracting with negative numbers

The Number Line is a brilliant tool for helping with this. Even secondary school pupils will benefit from using this prop – and should be encouraged to do so. Being able to visualise negative numbers as part of a sequence of all numbers helps comprehension.

> **Remember:** When subtracting we move to the *left* along the Number Line.

| -6 | -5 | -4 | -3 | -2 | -1 | 0 | 1 | 2 | 3 | 4 |

For example:

🎲 −2 − 3

Using the Number Line above, put your finger on −2 and as we're subtracting, jump along to the left 3 spaces.
Answer: −5

Now try

🎲 3 − 7

Put your finger on 3 and jump along to the left 7 spaces.
Answer: −4

And

🎲 −2 − 4

Start at −2 and jump along 4 spaces to the left.
Answer: −6

With the help of the Number Line, this is just as easy as subtracting any other numbers.

With smaller negative numbers, you just need to extend the Number Line further:

−12 −11 −10 −9 −8 −7 −6 −5 −4 −3 −2 −1 0 1 2 3 4 5 6 7 8 9 10 11

So,

$$-6 - 5 = -11$$
$$-4 - 8 = -12$$
$$9 - 17 = -8$$
$$-5 - 5 = -10$$
$$11 - 12 = -1$$
$$9 - 21 = -12$$

Around the time they reach secondary school, some children will be asked to think about what happens if they subtract a negative number from another number.

For example:

$$2 \text{ subtract } -3, \text{which can also look like}$$
$$2 - 3 \text{ or}$$
$$2 - (-3)$$

The key to solving this is to know that addition and subtraction are the **inverse** of each other (that is, subtraction will 'undo' addition and vice versa).

[Quick reminder: $6 + 2 = 8$
If we want to 'undo' adding on 2, we simply subtract 2 and we will be back to where we started (that is 6) $8 - 2 = 6$]

We *know* that: $\quad\quad\quad\quad\quad 5 + (-3) = 2$ (see page 103).
If we now want to 'undo' adding on (-3) we need to subtract (-3).
So, using the inverse rule: $\quad\quad 2 - (-3)$ must take us back to the **5** we started with.
Therefore: $\quad\quad\quad\quad\quad\quad 2 - (-3) = 5$

Here is another example:

$$8 \text{ subtract } -5 \quad \text{which can also look like}$$
$$8 - -5 \quad \text{or}$$
$$8 - (-5)$$
Answer: 13

It may seem bizarre, but subtracting a negative number appears like adding.

Often children will be told:

'Two negatives make a plus' or 'A minus plus a minus make a plus'.

For example:

$$4 - -3 = 4 + 3 = 7$$

This is true and helpful but children must understand the rule. '*Which two negatives?*' '*Where?*' and '*Why?*'

Importantly, the '*two negatives*' must be adjacent (touching or next to each other with nothing in between).

- So $8 - - 5$ is the same as $8 + 5$
  Answer: $8 - - 5 = 13$

But this 'rule' would *not* apply to a sum such as: $-8 - 5$

Yes, there are two negatives, but they are not adjacent.

- For $-8 - 5$ we can use the number line as normal: finger on -8, subtract 5 by jumping 5 places to the left.
  Answer: $-8 - 5 = -13$

You can see why some pupils get confused. So use '*two negatives make a plus*' with caution.

WHY DO TWO NEGATIVES MAKE A PLUS?

Here's a story your children might follow: Imagine a competition a bit like *Strictly Come Dancing* but a whole lot harsher. The contestants perform their routine and then 3 judges hold up their respective scores. The minimum score isn't 1, these judges are so cruel they can award negative scores too, in a range from a dismal and disappointing −5 up to an awe-inspiring 5.

Paul and Penny are next. They perform a beautiful sequence of footwork, twists and turns but with some technical faults.

The judges lift aloft their huge score cards.

Judge 1	Judge 2	Judge 3
1	4	−3

Adding up the scores, Paul and Penny achieve a total of:

$$1 \; + \; 4 \; + \; -3 \; \rightarrow \; 2$$

This total puts Paul and Penny in joint poll position on the score board. (In other words there is a draw.)

Now, in the case of a draw the 'golden rule' applies: The contestants' weakest score is disregarded.

So for Paul and Penny the **−3** is disregarded, that is, taken away from their original total.

Their new total is therefore:

2 take away (−3)

```
     ^                    ^
     |                    |
the original total   the weakest score
```

that is   **2 − −3**

But what is $2 - -3$?

Well another way to find their new total is to simply look at their original scores again and this time add them up without the −3. We can easily see the new total is 5 (as $1 + 4 = 5$).

So   **2 − −3 has to be 5, that is…**
                  2 − −3 = 5 (and is the same as 2 + 3)

Paul and Penny sweep to victory.

But more significantly… I hope this 'story' has shed a little light on the saying 'two negatives make a plus'. And that when you subtract a negative number, the original amount will increase.

Children may find it a little trickier when asked to consider what happens if we subtract a negative number from another negative number, but the same rules apply.

For example:

$$-2 - (-3)$$

Using the '*two negatives make a plus*' rule, the above could also be rewritten as:

$$-2 + 3 \quad \text{which equals } 1$$

Some more examples:

$$-1 - -6 = 5$$
$$-4 - -8 = 4$$
$$7 - -5 = 12$$
$$-8 - -7 = -1$$
$$9 - -3 = 12$$
$$-12 - -12 = 0$$
$$-12 - -9 = -3$$

So there it is. We have summed up Subtraction – and I hope this will be of help to you and your children throughout their primary and secondary school years.

The next chapter considers another of the four main operations of number: Multiplication.

4.
MULTIPLICATION

Multiplication is another of the key building blocks of our mathematical tower. It is among the most important, too, so this chapter sets out to give you confidence in helping your children understand its fundamental principles.

Mental methods of multiplication remain pertinent. Without them the written methods cannot be mastered. Long multiplication is still much the same as when we were at school but there are several 'new' staging posts along the way to help build understanding and confidence. More 'modern' methods for tackling long multiplication also exist and can be used in preference to the more traditional method.

UNDERSTANDING THE BASICS
(RECEPTION AND YEARS 1 AND 2, AGES 5–7)

Children are likely to hear and use many terms that all relate to multiplication: times, lots of, of, multiply, double, twice, multiplied by, multiples of, groups of, how many times as big (long, wide, tall...), times tables, product...

Multiplication is usually introduced to children a little later than addition and subtraction. As you will see below, children need to be able to add before they can be taught to multiply.

So, in the first couple of years at school, children may be introduced to: doubling very simple numbers (up to 5 + 5); counting up in twos, fives and tens; putting objects (blocks, counters, beads and so on) into groups and begin to appreciate the idea of 'lots of' something. 'Lots of' is something we might take for granted: '*I'll have 2 lots of eggs this week please.*'

Children will begin to understand the concept of 'lots of' by practising putting objects into groups. This will progress to children drawing these groups on paper.

For example:

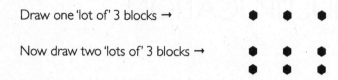

Draw one 'lot of' 3 blocks →

Now draw two 'lots of' 3 blocks →

Or

Draw one 'lot of' 2 blocks →

Now draw three 'lots of' 2 blocks →

All of this is in preparation for what comes next and the rest of this section relates to what most children will begin to experience from Year 2.

Repeated addition is one of the very first methods of multiplication children will be taught.

Children will need to have fully appreciated the idea of using 'lots of' a number.

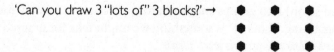

'Draw 2 "lots of" 3 blocks.' →

'How many blocks have you drawn?'

Children simply need to count the blocks or, better still, be encouraged to do the simple addition of 3 + 3.

'Can you draw 3 "lots of" 3 blocks?' →

'How many blocks have you drawn?'

Again, children may simply count the blocks but the idea is to progress from this and 'see' the simple addition of 3 + 3 + 3.

In time, children will learn to relate 'lots of' with adding. To find 2 'lots of' 3 you just add 3 two times (3 + 3). To find 3 'lots of' 3 you just add 3 three times (3 + 3 + 3). To find 4 'lots of' 3 you just add 3 four times (3 + 3 + 3 + 3).

And this is repeated addition.

When children are first introduced to the multiplication sign ('×') they will be explicitly taught that it is the same as saying 'lots of'.

So '5 × 3' is the same as '5 lots of 3', which is the same as 3 + 3 + 3 + 3 + 3, which equals 15.

SOMETIMES IT IS EASIER TO ADD THAN TO MULTIPLY

There is a simple connection between multiplication and addition: multiplication is shorthand for addition.

3 + 3 + 3 + 3 + 3 + 3 + 3 + 3 + 3 is the same as 9 'lots of' 3 which is the same as 9 × 3.

9 × 3 is a quicker or shorthand way of writing 3 + 3 + 3 + 3 + 3 + 3 + 3 + 3 + 3.

Understanding this connection can help build confidence and fluency in multiplication. This also explains why children need to be able to add before they can be expected to understand how to multiply.

The sum above can also be written as '3 × 5' or '3 lots of 5' (which is the same as 5 + 5 + 5 = 15).

One of the very first rules of multiplication that children need to know is that multiplication is **commutative.** They don't need to know this word, but right from the word go, they need to understand that multiplication can be done in any order. For example:

3 × 5 is exactly the same as 5 × 3 (that is 3 × 5 = 5 × 3)

A good way to convince children of this concept is for them to compare their drawings.

For 3 lots of 5 looks like this:

• • • • •
• • • • •   3 lots of 5 = 15
• • • • •

And 5 lots of 3 looks like this:

• • •
• • •
• • •   5 lots of 3 = 15
• • •
• • •

Each picture is the same as the other, simply turned on its side.

An **array** is simply a single picture used for demonstrating that multiplication is commutative, as explained above. It can be helpful for explaining why the rule is true, and for relating repeated addition with multiplication:

$4 \times 2 = 8$ (that is 4 columns each with 2 sweets)

$2 \times 4 = 8$ (that is 2 rows each with 4 sweets)

An array can be interpreted as either repeated addition or multiplication.

The array above can be described as:

$4 + 4$ or as 2×4 or as $2 + 2 + 2 + 2$ or as 4×2

The commutative rule is very important and can be a great help when children are trying to learn all their times tables. In fact, they really only ever need to learn half of their tables because if they know $5 \times 3 = 15$ they also know $3 \times 5 = 15$.

MULTIPLICATION SQUARE

×	1	2	3	4	5	6	7	8	9	10
1	1	2	3	4	5	6	7	8	9	10
2	2	4	6	8	10	12	14	16	18	20
3	3	6	9	12	15	18	21	24	27	30
4	4	8	12	16	20	24	28	32	36	40
5	5	10	15	20	25	30	35	40	45	50
6	6	12	18	24	30	36	42	48	54	60
7	7	14	21	28	35	42	49	56	63	70
8	8	16	24	32	40	48	56	64	72	80
9	9	18	27	36	45	54	63	72	81	90
10	10	20	30	40	50	60	70	80	90	100

The circled diagonal numbers split the square in half. Imagine this diagonal as a mirror line. Each side of the line is the reflection of the other. So 5 × 3 is a reflection of, and therefore exactly the same as, 3 × 5. Likewise 8 × 2 is the same as 2 × 8. And 9 × 6 is the same as 6 × 9. And...

The diagonal numbers are a special set of numbers called **Square Numbers** (that is 1 × 1 = 1, 2 × 2 = 4, 3 × 3 = 9, 4 × 4 = 16 and so on; see page 284).

The inverse idea is the second rule of multiplication that children need to become familiar with. Multiplication is the **inverse** of division, and vice versa. Again they don't need to know this word but they will need to appreciate that multiplication and division are opposite operations and each will undo or reverse the other.

Children will probably be introduced to the multiply ('×') sign in Year 2 at school and the equals ('=') sign earlier on. Children may start relating their 'maths chat' problems to multiplication sums, using these symbols. This is often referred to as writing number sentences.

Multiplication number sentences using the multiply ('×') and equals ('=') signs. Teachers will sometimes ask children to write **number sentences**. This can cause confusion for parents who are not familiar with this terminology, but it's a simple idea and a way of introducing children to writing down their workings.

Imagine Poppy had been doing some multiplication sums. She decided she wanted to give 5 marbles to each of her 3 sisters, Scarlet, Ruby and Tallulah. So she knew she needed 3 'lots of' 5. Poppy multiplied and realised she needed 15 marbles in total. '*Good work Poppy!*' says the teacher. '*Can you now record that as a number sentence?*'

Poppy writes: 3 × 5 = 15. '*That's it!*'

Using symbols: Teachers will use symbols such as ■ and ▲ to stand for numbers. This isn't tricky stuff really, but can throw you if you're not expecting them. The task is simply to work out which number each symbol must stand for.

For example:

8 × 5 = ■	Answer: ■ = 40
9 × ▲ = 18	Answer: ▲ = 2
⬢ × 2 = 12	Answer: ⬢ = 6
⬢ × ▲ = 100	Possible answers: ⬢ = 10 ▲ = 10
	or: ⬢ = 20 ▲ = 5
	or: ⬢ = 25 ▲ = 4
	or: ⬢ = 50 ▲ = 2

Much, much later the symbols will be replaced by letters and the eloquent language of algebra will be introduced.

Times tables will start to become part of your child's weekly routine at this stage. The *aim* will be for children to know, and be able to rapidly recall, their 2-, 5- and 10-times tables by the age of 7 (the end of Year 2).

It is definitely worth getting hold of one of those big posters of all the times tables (known as a **Multiplication Square**), often on sale in children's bookshops, libraries and catalogues, and sticking it on a wall somewhere. Put it somewhere your children can reach, so they can touch it and point at it, see the patterns and maybe even *learn* their tables!

You can also find posters of the **Hundred Square,** which is another very useful prop for explaining and understanding times tables. Your child can be shown how to start at 2 and then count on in twos for the 2-times table, start at 5 and then count on in fives for the 5-times table and start at 10 and then count on in tens for the 10-times table, spotting patterns as they go (see also page 18).

Reassuringly, learning times tables hasn't changed much from our day, so here's one area where we can confidently repeat what we heard in our own classrooms. One possible difference may be that children are sometimes asked to include '0 times' a number too, so a typical tables 'test' might start:

$$0 \times 2 = 0$$
$$1 \times 2 = 2$$
$$2 \times 2 = 4$$
$$............$$

and so on... up to $\qquad 10 \times 2 = 20$

Something dads and uncles in particular like to do is '*Give me 5, give me 10...*' with the children. This is great. It can really help young children start counting up in fives. One dad I know plays this game with his very sporty 5-year-old boy. They have great fun doing high fives, low fives and so on – '*Give me 15, give me 20...*'. It usually involves a lot of jumping, shouting, whooping and boy-type joyful energy.

The Number Line again!

Yes, the Number Line can also be used for multiplication. The most important point for your children to remember is that they *always* start at 0, and then jump along (or count up) the Number Line in whatever multiple they are dealing with.

So, for the 2-times table, finger on 0 and then jump along in twos, as shown below: 0, 2, 4, 6… This builds on a child's familiarity with the Number Line, helps show that multiplication is **repeated addition**, and makes the whole thing nice and easy.

Doubling

Prior to Year 2, children will be introduced to the idea of doubling and be expected to find the double of numbers up to at least 10 (that is 5 + 5). Clearly, to begin with they can use the fingers of both hands to help them with this, and these **number bonds** (see page 69) will soon become familiar. This will be extended in Year 2, so that children can quickly double all numbers up to at least 20 (that is up to 10 + 10).

Children will come to understand that doubling is exactly the same as multiplying by 2.

MOVING FORWARD
(YEARS 3 AND 4, AGES 7–9)

We've covered the basics of multiplication above. The same tools and rules are still used as your child progresses, generally using bigger numbers.

It's worth remembering at this point that the aim is for children to use mental ('in your head') methods wherever possible. But some calculations can't be done 'in your head' and this section introduces written methods, starting with informal **paper-and-pencil** recordings and building in stages towards more formal and concise methods.

As a rough guide, the aim will be that by the end of Year 4, most pupils can multiply a 2-digit number by a 1-digit number (for example 48 × 9).

But first we'll have a quick recap from the previous 'Understanding the Basics' section before introducing a few new concepts.

- Multiplication is **commutative** (that is, it can be done in any order so, for example, 8 × 7 is the same as 7 × 8).
- Multiplication is the **inverse** of division and vice versa (that is, one will 'undo' the other). For example, if **8** × 7 = 56, then 56 ÷ 7 will take you right back to the **8** you started with.

Following on from the above two rules, basically if you are happy that you know one fact, then in fact you know four. Bargain! For example, if you are happy that you know 12 × 8 = 96 then whether you realise it or not you now also know:

8 × 12 = 96 (using the commutative rule)
and
96 ÷ 12 = 8 (because division is the inverse of multiplication)
and also
96 ÷ 8 = 12

Now follow some new concepts and ideas that your children may be introduced to around now.

Repeated addition and the reverse

As we saw in the first section of this chapter, it's helpful and reassuring for children to understand that multiplication is the same as adding up the same number over and over again until we have the required amount of 'lots'. Although bigger numbers will be used at this stage, the rule remains the same.

For example:

'4 × 75'

Now, if the sum '4 × 75' looks a bit forbidding, we can add the numbers instead: as 4 × 75 is the same as 4 'lots of' 75 or 75 + 75 + 75 + 75.

$$4 \times 75 = 75 + 75 + 75 + 75 = 300$$

But being able to observe the reverse – that is, shortening addition sums into a simple multiplication – is often the appropriate aim for this age. For example, children should be able to 'see' 5 + 5 + 5 + 5 + 5 + 5 as 6 × 5.

In the *Third International Mathematics and Science Study*, Year 4 pupils were asked to write the addition sum '4 + 4 + 4 + 4 + 4 = 20' as a multiplication. Only 39 per cent of English pupils managed the correct response of '5 × 4 = 20'. (In contrast 90 per cent of pupils in the top-scoring country responded correctly.)

Scaling

Your children may be introduced to the principles of scaling up or scaling down, which will eventually lead to them understanding things like scale models, scale drawings, the scale on a map and so forth.

For example, children might be shown a blue tower made of 4 cubes and be asked to '*make one 3 times as high*' or '*find a ribbon twice as wide as the yellow one*' or '*look for a piece of string 4 times as long as this one*'. In this way you help your children perceive that multiplication can be applied to sizes and measures too.

Multiplying by 1

Multiplying a number by 1 always leaves a number unchanged. Children may understand this as one 'lot of' the number, such as one 'lot of' 6 sweets is 6 sweets.

$1 \times 6 = 6$ or
$6 \times 1 = 6$

Multiplying by 0

Multiplying a number by 0 will always give 0 as the result.

$6 \times 0 = 0$ or
$0 \times 6 = 0$

It's a rather abstract idea ('something times nothing is nothing!' or 'nothing times something is nothing!') until your children understand that 0 lots of something is nothing. For example, no 'lots of' sweets is no sweets!

'Of' means the same as 'times'

In maths '*of*' means the same as '*times*' as their use is interchangeable. For example: $\frac{1}{2} \times 8$ is the same as $\frac{1}{2}$ of 8. The second version might seem easier to do. **Answer: 4.** Likewise $\frac{1}{2}$ of $\frac{3}{4}$ is the same as $\frac{1}{2} \times \frac{3}{4}$, which might seem easier to tackle (especially when you have read Chapter 7, which deals with Fractions). **Answer: ⅜**

More multiplication laws

As your children work through maths exercises, you can continue to help reinforce the **commutative**, **associative** and **distributive**

laws (they will have been taught the principles, but usually not the names).

We have seen the **commutative law** above and in Chapter 2: Addition. It simply means that we can change the order of the numbers without affecting the outcome.

Examples of the commutative law:

$$7 \times 12 = 12 \times 7 = 84$$
$$3 \times 5 = 5 \times 3 = 15$$
$$10 \times 8 = 8 \times 10 = 80$$

The **associative law** was created to resolve the issue of what to do if we have three or more numbers to multiply – which numbers do we multiply first? For example: $4 \times 5 \times 2$. Do we do the 4×5 first or the 5×2? The answer is that it doesn't matter, and that is exactly what the associative law illustrates.

The associative law allows you to use and move brackets (parentheses) to show that it doesn't matter.

For example:

$$4 \times 5 \times 2 \quad = 4 \times (5 \times 2) \qquad \text{or} \quad = (4 \times 5) \times 2$$
$$= 4 \times 10 \qquad\qquad\qquad = 20 \times 2$$
$$= 40 \qquad\qquad\qquad\qquad = 40$$

The associative law can also be used to simplify multiplication. Consider:

$$5 \times 24$$

The 24 can also be thought of as 6 'lots of' 4 (6×4). The multiplication can now be rewritten as:

$$5 \times 6 \times 4$$

Using the associative law:

$$5 \times 6 \times 4 \quad = \quad (5 \times 6) \times 4$$
$$= \quad 30 \times 4$$
$$= \quad 120$$

Examples of associative law:

$$6 \times \ 5 \times 2 \ = \ 6 \times (5 \times \ 2) \quad or = (\ 6 \times 5) \times \ 2$$
$$10 \times \ 8 \times 7 \ = 10 \times (8 \times \ 7) \quad or = (10 \times 8) \times \ 7$$
$$4 \times 60 \quad = \ 4 \times (6 \times 10) \quad or = (\ 4 \times 6) \times 10$$

In the case of the **distributive law**, it might be easiest to start with an example to explain this rule. Let's look at:

$$4 \times (2 + 3)$$

One way to get to the answer is to do the 'bit' in the brackets first. This then gives us:

$$4 \times 5 \quad \text{which equals } 20$$

But another way to get to the answer is to use the distributive law. Using the distributive law, we can 'distribute' the '4' over each 'bit' in the brackets to arrive at the same answer. I'll show you what I mean:

$$4 \times (2 + 3) \ = \ (4 \times 2) + (4 \times 3) \ = \ 8 + 12 \ = \ 20$$

The '4' has been multiplied across each 'bit' or term within the original brackets.

Examples of distributive law:

$$3 \times (10 + \ 4) = (\ 3 \times 10) + (3 \times \ 4) = 30 + 12 = \ 42$$
$$8 \times (\ 6 + 12) = (\ 8 \times \ 6) + (8 \times 12) = 48 + 96 = 144$$
$$5 \times (11 - \ 2) = (\ 5 \times 11) - (5 \times \ 2) = 55 - 10 = \ 45$$
$$(10 - 2) \times 7 \ = (10 \times \ 7) - (2 \times \ 7) = 70 - 14 = \ 56$$

The distributive law can also be used in 'reverse':

$$(12 \times 3) + (12 \times 7) = 12 \times (3 + 7) = 12 \times 10 = 120$$

The '3' and the '7' are both being multiplied by 12, so this is the same as 12 'lots of' '3 + 7' .

More times tables

By the end of Year 2 your children will have had some success at mastering their 2-, 5- and 10-times tables. In Year 3 they will also be expected to learn their 3-, 4- and 6-times tables and then to know all their tables up to 10 × 10 by the end of Year 4.

Some times tables are really easy to learn. The obvious ones are the 5- and 10-times tables. Think of *'give me 5, give me 10...'* as a starting point for the 5-times table. Hopefully, your children will also find the 2-times table relatively easy. At school they may well practise *'say one, miss one'* from the Number Line. For example: *'0...,2...,4...,6... and so on'*. This is probably the way your children were also taught their **even** numbers (**even** numbers are the numbers that can be divided by two exactly, hence they are exactly the same as the numbers in the 2-times table).

For many children, little ditties can also help to get the ball rolling:

'2, 4, 6, 8, who do we appreciate!'
'3, 6, 9, you're all mine!'

It helps if you and your children can make up your own – usually the sillier the better.

For the **9-times table,** try this trick, which all children seem to like:

- Hold up all 10 fingers in front of you (thumbs and fingers are all the same in this trick).
- Imagine you wanted to know 7 × 9.
- Bend the seventh finger down (*working from left to right*).
- Now, to the left of the bent finger you will have 6 fingers

(representing 6 tens) and to the right of the bent finger you
will have 3 fingers (representing 3 units).

- **Answer:** 63

Now imagine you wanted to know 3 × 9

- Bend the third finger down (*working from left to right*).
- Now, to the left of the bent finger you will have 2 fingers and
  to the right of the bent finger you will have 7 fingers.
- **Answer:** 27

FLUENCY IN TIMES TABLES IS IMPORTANT

Why do I think that? Well, without it, I think children will forever
face unnecessary struggles. If children are being taught about area,
for example, and they are learning to calculate the area of a
rectangle, the lesson being taught is that the area of a rectangle
equals the width multiplied by the length. Children who are not
fluent with times tables will labour over the multiplication process,
instead of learning the new lesson. And being a dab hand at division depends almost entirely on being triumphant with your
times tables.

But, as always, the exception proves the rule. Over the years
I have taught some exceptionally talented mathematicians who
did not know their times tables. So, if your children are really
struggling with the rote-learning but seem to grasp the underlying mathematics please don't despair.

At this stage, children will still be expected to do most of their
multiplication sums in their head, so here are some top tips that
you can offer to help them. These **mental multiplication tips** are
very much encouraged in today's classroom.

To multiply by 4, double and double again.

- For example: 4 × 18 is the same as double (double 18), which
  is double 36 which equals 72.

This method is really very popular in schools now – so it's one that is probably worth encouraging your children to use at home.

- $26 \times 4 = $ (26 doubled and then doubled again) $= 26 \times 2 \times 2 = 52 \times 2 = 104$
- $4 \times 145 = 2 \times 2 \times 145 = 2 \times 290 = 580$

Doubling can also help with multiples of 25:

1 lot of 25 is 25 so...	$1 \times 25 = 25$
2 lots of 25 must be 50 so...	$2 \times 25 = 50$
4 lots of 25 must be double 50 (that is 100) so...	$4 \times 25 = 100$
8 lots of 25 must be double 100 (that is 200) so...	$8 \times 25 = 200$
16 lots of 25 must be double 200 (that is 400)	$16 \times 25 = 400$

To multiply by 5, multiply by 10 and halve that answer. For example:

- 5×32 is the same as half of (10×32) or half of 320, which equals 160.

To multiply by 20, multiply by 10 and then double that answer.

- For example: 20×32 is the same as **double (10×32)** or **double of 320**, which equals 640.

Work out the 8-times table by doubling the 4-times table facts:

$1 \times 4 = 4$	so	$1 \times 8 = 8$	
$2 \times 4 = 8$	so	$2 \times 8 = 16$	
$3 \times 4 = 12$	so	$3 \times 8 = 24$	
$4 \times 4 = 16$	so	$4 \times 8 = 32$	

Work out the 6-times table by adding 2-times table facts to the 4-times table facts:

$1 \times 4 = 4$	and	$1 \times 2 = 2$	so	$1 \times 6 = 6$
$2 \times 4 = 8$	and	$2 \times 2 = 4$		$2 \times 6 = 12$

$$3 \times 4 = 12 \quad \text{and} \quad 3 \times 2 = 6 \qquad 3 \times 6 = 18$$
$$4 \times 4 = 16 \quad \text{and} \quad 4 \times 2 = 8 \qquad 4 \times 6 = 24$$
$$\ldots \qquad\qquad\qquad \ldots \qquad\qquad\qquad \ldots\ldots$$

To multiply a number by 11, multiply by 10 and then add the number:

$$14 \times 11 = (14 \times 10) + 14$$
$$= (140 + 14)$$
$$= 154$$
$$21 \times 11 = (21 \times 10) + 21$$
$$= (210 + 21)$$
$$= 231$$

To multiply a number by 9, multiply by 10 and then subtract the number:

$$14 \times 9 = (14 \times 10) - 14$$
$$= (140 - 14)$$
$$= 126$$
$$21 \times 9 = (21 \times 10) - 21$$
$$= (210 - 21)$$
$$= 189$$

Use the idea of partitioning and the distributive rule to help make multiplication easier.

For example:

- $23 \times 3 = (20 \times 3) + (3 \times 3) = 60 + 9 = 69$
- Double 43 = double 40 add double 3 = 80 + 6 = 86

Multiplying by 10 or 100 is a very important skill for your children to master.

For example:

$$4 \times 10 = \quad 40 \qquad\qquad 3 \times 100 = \quad 300$$
$$32 \times 10 = \quad 320 \qquad\qquad 56 \times 100 = \quad 5600$$
$$763 \times 10 = 7630 \qquad\qquad 928 \times 100 = 92800$$

Something that we were probably all taught at some point was *'just add a nought if you are multiplying by 10, and add two noughts if you are multiplying by 100'*. It's a shortcut we probably still use ourselves, so is it okay to tell your children this rule?

The problem with this is that children taught this rule parrot fashion, without an underlying understanding of why it is true, may come unstuck when things get more complicated. It **doesn't** work with decimal numbers, for example, but only with whole numbers (see also page 31).

Chapter 1 provides a fuller discussion of **place value** (which is what this is) but meanwhile I think the common sense approach is to help your children practise doing a lot of multiplication by 10 and 100 and then see if they *notice* any pattern. If they do, then they might start using this 'rule' for themselves anyway.

Related multiplication facts

As we have seen in other chapters, using **related facts** can be a big help. This just means using a simpler sum to help with a harder one.

For example, asked to find '30 × 7' we can use '3 × 7' to help:

$$3 \times 7 = 21 \quad \text{so} \quad 30 \times 7 = 210$$

Why?

Well	30 × 7
is the same as	(3 × 10) × 7
And as multiplication can be done	
in any order, this can be rewritten as	3 × 7 × 10,
which equals	21 × 10,
which is	210

And similarly, using the same related fact:

$$3 \times 70 = 210$$
...and $\quad 30 \times 70 = 2100$
...and $\quad 300 \times 7 = 2100$
...and $\quad 3 \times 700 = 2100$ ...and so on.

Another typical example: 40 × 8.

Using the *related fact* that 4 × 8 = 32 would lead them to know that 40 × 8 = 320.

Once again, the explanation of why this is so is:

First, 40 × 8 can be rewritten as   (4 × 10) × 8
Multiplication can be done in any order,
so the sum could be rewritten again as   (4 × 8) × 10
Children know that 4 × 8 is 32, so now
all they have to do is multiply by 10 to
get the final answer.   32 × 10 = 320

Partitioning (see page 74) is also used to help with mental multiplication. For example, to do a sum such as 17 × 3 in your head, the 17 can first be partitioned into its component **tens** and **units**: 17 → 10 + 7.

- The 10 and the 7 can then be multiplied separately by 3.
- 10 × 3 is 30 and 7 × 3 is 21.

These partial sums are then added, or **recombined**, to give the final answer:

30 + 21 = 51

In summary:

17 × 3 → (10 + 7) × 3 → (10 × 3) + (7 × 3)
= 30 + 21 = 51

Partitioning is certainly helpful for mental multiplication, but it is also useful for the written methods of multiplication outlined below.

As parents, we'll begin to see **informal written methods** (sometimes referred to as 'pencil-and-paper' methods') being introduced at this stage. Writing things down is intended simply to build on their understanding so far, and **standard written methods** (see below) follow some time later.

The written methods are built up in stages, with the intention of helping children progress to the next stage when they feel confident to do so. The end point (for some – although not all pupils will reach this stage) is the conventional long multiplication method, with which you may be familiar from your own school days.

For children to use these written methods with ease, they must be able to build on everything they have been taught so far and have a very secure understanding of place value (see pages 16 and 30).

In summary: to multiply with ease children must be confident with the following:
- all times tables facts up to 10 × 10;
- understand how to partition; for example: 39 is the same as 30 + 9;
- work out sums such as 30 × 7 by using the related fact that 3 × 7 = 21 and their knowledge of place value, so that 30 × 7 = 210 (see above);
- add together whole numbers – either mentally or using the column method (see page 93).

The stages...

Stage 1: Partitioning

As we saw above, partitioning is a very helpful way to multiply.

An early written method simply records the steps taken. An informal written recording of 47 × 3 might look like this:

```
        47
40   +      7      ×  3
120  +      21     = 141
```

Or this sum may be written like this:

$$47 \times 3 \quad = \quad (40 + 7) \times 3$$

Now we can rewrite this (using the distributive law, see page 163) as:

$$
\begin{aligned}
47 \times 3 \quad &= \quad (40 \times 3) + (7 \times 3) \\
&= \quad 120 + 21 \\
&= \quad 141
\end{aligned}
$$

Another example:

$$
\begin{aligned}
53 \times 8 \quad &= \quad (50 + 3) \times 8 \\
&= \quad (50 \times 8) + (3 \times 8) \\
&= \quad 400 + 24 \\
&= \quad 424
\end{aligned}
$$

Stage 2: Multiplication grid or box method

This is a way of making long multiplication easy to understand and easy to do. For most parents, this is new and unfamiliar – but it is a really simple method for multiplying.

Using the grid method, 34×8 would look like this:

×	8
30	$30 \times 8 = 240$
4	$4 \times 8 = 32$

Answer: 272

The 34 has been partitioned into 30 and 4 and each part multiplied by 8. The 240 and 32 are then added together to give the answer 272. It may be less efficient than 'our' old way of doing it, but children are less likely to make mistakes, and so the method is considered more 'reliable'. Some children will continue to use this method, or a variant of it, well into secondary school and beyond, never mastering (and in fact not needing to use) the traditional method of long multiplication. Others will be introduced to traditional long multiplication in Years 5 and 6 and this may then, in time, completely replace their 'need' for the grid method. Others will use a combination of the two throughout their schooldays.

Here is another example to show how the grid can work for bigger numbers.

26 × 32:

×	30	2
20	20 × 30 = 600	20 × 2 = 40
6	6 × 30 = 180	6 × 2 = 12

Answer: 832

The 26 has been partitioned into 20 and 6; the 32 into 30 and 2. There are now *four* multiplications to do mentally, using related facts to help where necessary (for example, 2 × 3 = 6 so 20 × 30 = 600).

The 600, 180, 40 and 12 are then added, either mentally or using the column method (see also page 93) to give the answer 832.

Why does this grid method prevent mistakes and simplify multiplication for children? It does so by clearly breaking the multiplication and addition into two distinct steps. In this way, it demystifies the sum and helps make clear the logical steps the children need to follow. Our 'traditional' method of long multiplication is just a bit less step-by-step, and that's why this grid method is now taught at this stage.

Parents regularly bemoan the fact that they are out of practice with long multiplication, just as their children are asking for help with it. With a calculator on every mobile phone and a spreadsheet package on every laptop, this is not surprising. And so, learning a multiplication method that relies more on logic and less on memory can be quite useful for adults too (see also page 183).

Stage 3: Expanded short multiplication

This method follows on from the grid method (outlined above) and can be seen as a staging post on the way to the more traditional or conventional procedure.

Short multiplication is a term with which you may not be familiar. It simply refers to situations where one number is being multiplied by a single digit. For example:

- 34 × 8 is a short multiplication; so too is 427 × 9.

For a sum like **34 × 8**, children will first be asked to *approximate* the answer – in other words, work out roughly what the answer will be (see also page 41).

In this case, **34 × 8** is less than **34 × 10** which equals **340**. This can then be used to check the 'reasonableness' (in particular the magnitude or size) of the calculated answer.

Now for the sum itself. This is now recorded in a column format in one of two ways:

Either:			*Or:*	
30 + 4			34	
× 8			× 8	
———			——	
240	30 × 8		240	
32	4 × 8		32	
———			——	
272			272	

As you can see, this is along very similar lines to the grid method, but the presentation is different. This time the partitioning of 34 is less explicit and in the second version the partitioning is now done 'in your head'. Then we continue as before and do '*30 times 8*' followed by '*4 times 8*'. Both these partial answers are recorded and then added to give the final answer of 272.

This answer is then compared with the approximated answer to check that it is a reasonable outcome.

Once again, as discussed in other chapters, it is essential that the **units** line up under **units**, **tens** under **tens**, **hundreds** under **hundreds** and so on.

Here is another example: 47 × 8

First, we will approximate the answer. Well, 47 × 8 will be less than 50 × 8 and using the related fact that 5 × 8 = 40 we can deduce that 50 × 8 = 400. We will expect our final answer to be less than, but in the region of, 400.

$$
\begin{array}{r}
47 \\
\times\ \ 8 \\
\hline
\end{array}
$$

40 × 8	320
7 × 8	56
	———
	376

The final answer of 376 seems very reasonable in comparison with our approximation.

Stage 4: Standard short multiplication

Having laid the foundations, traditional multiplication methods *may* be introduced to *some* pupils at this point. The focus will still be on short multiplication (that is, one number multiplied by a single digit). Long multiplication will follow later (see page 176).

For example 47 × 8 would look like this:

$$
\begin{array}{r}
47 \\
\times\ \ 8 \\
\hline
{}^{5}\ \ \\
376 \\
\hline
\end{array}
$$

Now instead of doing 40 × 8 followed by 7 × 8 – as in the previous stages – we would first do 8 × 7 and then 8 × 40. The multiplications are identical, just the order in which they are performed is different. And the reasoning would go like this:

- '8 × 7 is 56, so I'll carry the 50 and write down the 6.
  And 8 × 40 is 320, adding this to the "carried" 50 makes 370.
  So 370 + 6 makes 376.'

The 'carried' 50 is shown by jotting a 5 in the tens column. Alternatively the 'carried' amount can be remembered in your head.

This may be very similar to what you would have done (and maybe still do), only 'yours' may have sounded more like this:

- '8 × 7 is 56, so I'll carry the 5 and write down the 6.
  And 8 × 4 is 32, added to the carried 5 makes 37.'

The first version is technically more correct and mirrors the grid and expanded methods. It is important for children to make the connections between the methods they have been taught so far and this new method.

But the second version is quick and concise and, having built the fundamental understanding first, this may eventually be the appropriate goal for some pupils.

There are now just two more stages…

- Expanded long multiplication
- Standard long multiplication

And these follow on in the next section as they are aimed at slightly older children.

TO THE TOP
(YEARS 5 AND 6 PLUS, AGES 9–12 PLUS)

As your children continue on their mathematical journey, they will find that all the rules and methods for multiplication that we've covered in the sections above are still in daily use. Expect bigger and smaller numbers and more complex questions, but the same principles are at work. In this section, we familiarise ourselves with some more terms and methods, and make the move into decimals and negative (minus) numbers.

But first we will complete the stages in developing an effective method for long multiplication – the first four stages of which are outlined in the previous section. As a rough guide, the aim is for children to be equipped with the necessary tools to be able to multiply a 2-digit number by a 2-digit number by the end of Year 5 (for example, 83 × 62), and to multiply a 3-digit number by a 2-digit number by the end of Year 6 (for example, 783 × 42).

Now follow the final two stages...

Stage 5: Expanded long multiplication

Just as for 'expanded short multiplication', covered in the previous section, expanded long multiplication follows on from the **grid method** (see page 171).

Long multiplication is a term with which you *may* be familiar but what does it actually mean? It means that both numbers being multiplied are bigger than single digits. For example 34 × 28 is an example of long multiplication. So too are 427 × 89 and 720 × 345.

- 'Expanded long multiplication' can be seen as a staging post on the way to traditional or 'standard long multiplication'.

Before starting to calculate a sum like 26 × 32, children will first be asked to *approximate* the answer, or work out roughly what the answer will be (see also page 41).

In this case 26 × 32 is approximately 30 × 30, which equals 900. This can then be used to check the 'reasonableness' (in particular the magnitude or size) of the calculated answer.

Now for the sum itself:

$$\begin{array}{r} 26 \\ \times\ 32 \\ \hline \end{array}$$

20 × 30	600
6 × 30	180
20 × 2	40
6 × 2	12
	832

As you can see, this is along very similar lines to the grid method (see page 171), with the presentation in columns instead of boxes. This time the partitioning of 26 and 32 is less explicit and is now done 'in your head'. We then continue as before and do *20 × 30* followed by *6 × 30*, then *20 × 2* and *6 × 2*. The partial answers are recorded and then added to give the final answer of 832.

A common mistake is for children to 'forget' to do all four of the multiplications. Typically they might only do 20 × 30 and 6 × 2. (I can't tell you how many times I've seen this mistake!) To resolve this, take your children 'back' a step to the grid method (see page 171) and show them the importance of filling in every box in the grid. Only move on again when they are confident enough (or drilled enough) to remember all the stages of the multiplication consistently.

And as the numbers involved get bigger, there are more multiplications to do. For example:

$$\begin{array}{r} 364 \\ \times\ 28 \\ \hline \end{array}$$

300 × 20	6000
60 × 20	1200
4 × 20	80
300 × 8	2400
60 × 8	480
4 × 8	32
	10192

Once again, as discussed earlier and in other chapters, it is essential that the **units** line up under **units**, **tens** under **tens**, **hundreds** under **hundreds** and so on.

From this point, a version of long multiplication *may* be introduced. For example:

$$
\begin{array}{r}
64 \\
\times\ 28 \\
\hline
\end{array}
$$

64 × 20	1280
64 × 8	512

$$
\begin{array}{r}
\hline
1792
\end{array}
$$

This time only the 28 has been partitioned so more complex mental multiplication is required. This may, or may not, suit your children. If not, leave it and stick with the version above.

Stage 6: Standard long multiplication

Some children *may* then (finally, I hear you cry!) be shown something very similar to 'our' way of doing long multiplication, or what we might call the 'traditional' or conventional method.

You and I probably did our sums at school like this:

$$
\begin{array}{r}
64 \\
\times\ 28 \\
\hline
5\overset{3}{1}2 \\
1280 \\
\hline
1792
\end{array}
$$

And you probably said to yourself:

'*8 × 4 is 32, so I'll carry the 3* and write down the 2 . And 8 × 6 is 48, which I'll add to the 3 that I carried to make 51. Now I'll start multiplying with the 2, but because it is not really 2 but 20, I need to* *write down a 0 first. So 2 × 4 is 8 and 2 × 6 is 12. Now all I need to do is add up the two lines. Hurrah!'*

Today's children will probably do it *ever so* slightly differently (sorry, it's never *exactly* like we did it):

$$
\begin{array}{r}
64 \\
\times\ 28 \\
\hline
1280 \\
5\overset{3}{1}2 \\
\hline
1792
\end{array}
$$

This is really very similar to 'our' way, the only difference being that they would start multiplying with the 20 first.

So Freddie says to himself:

'I'll start multiplying with the 2, but because it is not really 2 but 20, I need to write down a 0 first. So 2 × 4 is 8 and 2 × 6 is 12.'

(Some children cross out the 2 at this stage to remind them that they have finished with it.)

'Now for the next line: 8 × 4 is 32, so I'll carry the 3* and write down the 2. And 8 × 6 is 48, which added to the 3 that I carried makes 51.

'Now all I need to do is add up the 2 lines. Job done!'

* Note that when multiplying, the carried digits can either be jotted down – as with the '3' in this example – or alternatively can be carried mentally (that is just remembered in your head). If they are jotted down – as in the example above – you must be careful NOT to include them whilst adding up.

Here is another example:

$$
\begin{array}{r}
168 \\
\times\ 42 \\
\hline
6\overset{2\ 3}{7}20 \\
3\overset{1\ 1}{3}6 \\
\hline
7056
\end{array}
$$

And James says to himself:

'I'll start multiplying with the 4, but because it is not really 4 but 40, I need to write down a 0 first. So 4 × 8 is 32 so I'll carry the 3 and write down the 2. Next 4 × 6 is 24 which added to the 3 that I carried makes 27. I'll carry the 2 and write down the 7. And 4 × 1 is 4 which added to the 2 that I carried makes 6, which I'll write down.

'Now for the next line: 2 × 8 is 16, so I'll carry the 1 and write down the 6. And 2 × 6 is 12, which added to the 1 that I carried makes 13. So I'll carry the 1 and write down the 3. And 2 × 1 is 2, which added to the 1 that I carried makes 3, which I'll write down.

'Now all I need to do is add up the 2 lines.

Answer: 7056.'

That is the end of the final stage and shows how written multiplications may be illustrated in your child's school.

However, not all children and not all schools use this method. Quite a lot of children get into a mess when attempting to use it, so it is understandable that some schools steer clear of these stormy seas! It can simply seem too abstract to remember and combine the steps, especially when not in daily use.

So the more **informal multiplication methods** such as the **grid method** are still very much in use at this stage – but with bigger numbers and so with bigger boxes.

There are some more grid methods (more of which shortly) that are so efficient that there really is no need to introduce 'old' or conventional column methods.

Using the grid method, 43 × 16 would look like this:

×	10	6
40	40 × 10 = 400	40 × 6 = 240
3	3 × 10 = 30	3 × 6 = 18

The 43 has been partitioned into 40 and 3 and the 16 has been partitioned into 10 and 6. The partial answers of 400 and 30 and 240 and 18 are then added together:

400
240
 30
 18
___
688

Using the grid method in a slightly more abbreviated way, 64 × 28 would look like this:

×	20	8
60	1200	480
4	80	32

The 64 has been partitioned into 60 and 4 and the 28 has been partitioned into 20 and 8. We add together the partial answers, to give the final answer:

1200
 480
  80
  32
____
1792

The grids can be drawn to any size to suit any sum:

24 × 8

×	20	4
8	160	32

Answer: 160 + 32 = 192

15×3162

×	10	5
3000	30000	15000
100	1000	500
60	600	300
2	20	10

Answer:
30000
1000
600
20
15000
500
300
10
_____
47430

53×234

×	50	3
200	10000	600
30	1500	90
4	200	12

Answer:
10000
1500
200
600
90
12
_____
12402

Let's remember: multiplication is **commutative**. In other words 24×8 is the same as 8×24. Likewise 15×3162 is the same as 3162×15, 53×234 the same as 234×53, and so on. So it doesn't matter how the boxes are drawn or which number goes along the top and which down the side. The answer will be the same whichever way you do it. The order in which you add up the numbers is also irrelevant to the answer.

Now – as I mentioned above – I will show you a method of long multiplication that can completely replace the need for the traditional method, and which many schools prefer to use as their standard.

Unlike the grid or box method shown so far, which can be cumbersome and time consuming – especially when big numbers are involved – this one is surprisingly succinct.

Napier's Method (or 'Napier's Bones' or the 'Lattice Method')

Some children find setting up the grid or lattice a little tricky to begin with. This can easily be managed by giving children pre-drawn grids. Most children master the art of grid drawing very quickly.

Here an example is worked through in detail:

672 × 39

First we draw a rectangular grid or box. The grid needs to be big enough for all the digits (that is, have the right number of rows and columns to accommodate the digits in the multiplication).

In this example we need a grid with three columns and two rows. 672 is written across the top, 39 down the right-hand side.

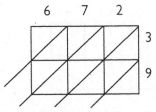

Now we need to draw the very important diagonal lines.

Note how the diagonal lines 'cut' or split each small square (or 'cell' to be technically correct) exactly in half. The diagonals must also 'run' from top right to bottom left – and should extend a little beyond the grid.

Now we are ready to start multiplying.

Each digit across the top is multiplied with each digit down the right-hand side.

So the multiplications are:

6 × 3
6 × 9
7 × 3
7 × 9
2 × 3
2 × 9

Now the answer to 6 × 3, that is 18, is placed in the square highlighted below:

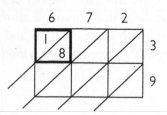

Importantly the **tens** ('1' in this case) are recorded above the diagonal line and the **units** ('8' in this case) are recorded below the diagonal line.

The rest of the multiplications are now shown:

6 × 9 = 54
7 × 3 = 21
7 × 9 = 63
2 × 3 = 6*
2 × 9 = 18

* Note: For 2 × 3 there are no **tens**, only **units**. This must be recorded as 06 as shown in the square highlighted.

Now we are ready to add up.

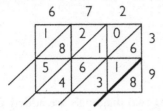

Starting at the bottom right-hand corner there is just one digit beneath the first diagonal line. This is recorded outside the grid as shown:

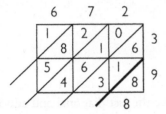

And now the digits between this diagonal line and the next are added: 6 + 1 + 3 = 10. So the 0 (**units**) is recorded outside the grid and the 1 (**ten**) is carried to the left as shown:

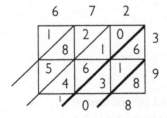

Now the digits between the next two diagonals are added, remembering to add on the carried digit: 0 + 1 + 6 + 4 + 1 = 12

The 2 (**units**) is recorded outside the grid and the 1 (**ten**) carried as shown:

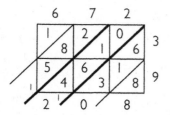

Now the digits between the next two diagonals are added again remembering to include the carried digit: 2 + 8 + 5 + 1 = 16

The 6 is recorded and the 1 carried as shown:

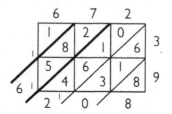

Finally we look at the digit above the last diagonal and add it to any carried digits.

In this case: 1 + 1 = 2

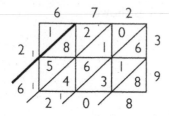

And now all we need to do is read off the answer.

Simply start at the top left of the grid....

Answer: 26208

Here another example, 64 × 28, would look like this:

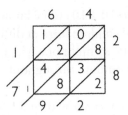

Answer: 1792

And here is another example: 387 × 42

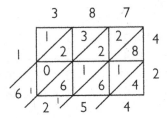

Answer: 16254

Very quick, very easy! And also very clever as the diagonals 'hold' the place value. So instead of having to do 300 × 40 for example, as in the above example, we do 3 × 4 but the answer of 12 is actually 'held' in the position of 12000 (which is the correct answer to 300 × 40) by the positioning of the diagonal 'pathways'.

This method was invented by John Napier (a Scot from Edinburgh) over 400 years ago, who was bored by all the long multiplication he had to do. As no one had yet had the foresight to create the calculator, he decided to come up with a better way. Pupils who go on to study maths at higher levels will come across logarithms, of which Napier's Bones were the precursor.

For what it's worth, in my experience Napier's Bones works really well for most children in the early years of secondary school. Once they've mastered drawing the grids, the rest is easy. It relies only on multiplying single digit numbers. (In other words: if they know their times tables they'll find it easy-peasy.)

More times tables

The aim is for all the times tables (up to 10×10) to be known by heart by the end of Year 4. In reality, all children learn at different speeds, and a bit of revision never hurts, so more practice, familiarity and plain old rote-learning will continue throughout Years 5, 6, 7 and beyond.

In an ideal world, your children will also know their 11-times table and 12-times table before starting secondary school. The 11-times table is really easy to learn as it starts with a very simple pattern: 11, 22, 33, 44, 55, 66, 77, 88, 99 and 110.

And one easy way to work out the 12-times table is to add the 2-times table to the 10-times table:

$1 \times 10 = 10$	and	$1 \times 2 = 2$	so	$1 \times 12 = 12$
$2 \times 10 = 20$	and	$2 \times 2 = 4$		$2 \times 12 = 24$
$3 \times 10 = 30$	and	$3 \times 2 = 6$		$3 \times 12 = 36$
$4 \times 10 = 40$	and	$4 \times 2 = 8$		$4 \times 12 = 48$
...		...		....

As well as practising their times tables on a regular basis, children will still be expected to do a lot of multiplication mentally ('in their heads'). This is a really good thing. A mental calculation can give you a rough idea of what to expect if you then go on to do a more detailed written multiplication and/or if you decide to punch some buttons on a calculator.

Below are a few principles that are worth stressing and which may be particularly helpful when doing mental multiplication.

Multiplying makes it bigger...

...when a (positive) number is multiplied by a number *bigger* than 1.

For example

$$3 \times 1.5 = 4.5$$

But...

Multiplying makes it smaller...
...when a (positive) number is multiplied by a number *smaller* than 1.

For example

$$3 \times 0.5 = 1.5$$
or $$3 \times \tfrac{1}{10} = \tfrac{3}{10}$$
or $$3 \times -2 = -6$$

(See also pages 197, 203 and 329).

The skill of **multiplying by 10 and 100** will have been introduced by now (see also pages 32 and 42). It is now extended to multiplying by higher powers of 10, such as 1000, and multiplying with decimal numbers. For example:

$4 \times 10 = 40$	$3 \times 100 = 300$	$7 \times 1000 = 7000$
$32 \times 10 = 320$	$56 \times 100 = 5600$	$49 \times 1000 = 49000$
$5.2 \times 10 = 52$	$9.8 \times 100 = 980$	$3.5 \times 1000 = 3500$

In the previous section, we asked whether it is okay to teach your children to '*just add a nought if you are multiplying by 10, add two noughts if you are multiplying by 100, or add three noughts if you are multiplying by 1000*'. The general answer is No! When multiplying with decimal numbers, for example 9.8 × 100, if children have just learnt parrot fashion to add a nought here or there with no understanding why they are doing it they will make all sorts of mistakes.

Common **WRONG** answers are:

$$9.8 \times 100 = 9.800$$
$$9.8 \times 100 = 900.8$$
$$9.8 \times 100 = 90.80$$
$$9.8 \times 100 = 90.8$$

The **CORRECT** answer is:

$$9.8 \times 100 = 980 \quad \text{or} \quad 9.8 \times 100 = 980.0$$

We need to think about our 'place value' again and see the number as:

H	T	U	.	t
		9	.	8

Now, multiplying by 100 means each digit becomes 100 times bigger, and so the answer is:

H	T	U	.	t
9	8	0	.	0

Another example:

4.236 × 100

H	T	U	.	t	h	th
		4	.	2	3	6

Answer:

H	T	U	.	t	h	th
4	2	3	.	6	0	0

So 4.236 × 100 = 423.6

You may have been told to: '*Move the decimal point one place to the right if multiplying by 10, two places if multiplying by 100 and three places if multiplying by 1000.*'

This is probably what you and I do, and we'd get the right answer. But is this okay to say to our children? To be technically correct, we should echo what's generally taught in classrooms today. That is:

- The *digits* move, but the decimal point stays firmly in its place.
- When multiplying, the *digits* move to the left (one place if multiplying by 10, two places if multiplying by 100, and three if multiplying by 1000 and so on).

This is definitely the more correct explanation, though it doesn't have the same succinct ring to it!

Again, the best approach is probably to practise doing a lot of multiplication of decimal numbers by 10, 100 and 1000, and see if your children *notice* any pattern. If they do, then they might start working out a 'rule' for themselves.

Calculating with hundreds and thousands does result in many common incorrect answers, even with relatively simple multiplications.

Look at:

$$3000 \times 200$$

Typical *incorrect* responses include:

6000 or 60 000

To get to the correct answer we could rewrite the sum as:

$$3000 \times \quad 2 \times 100$$
$$\rightarrow \quad 6000 \times 100$$

And then use a table similar to the ones above:

$$6000 \times 100$$

Th	H	T	U
6	0	0	0

Answer (moving each digit 2 places to the right):

H of Th	T of Th	Th	H	T	U
6	0	0	0	0	0

So 3000 × 200 = 600 000

More mental arithmetic strategies

Following on from the previous section, we saw how children are encouraged to find quick-and-easy ways to calculate answers in their head.

In the examples below, each number is broken down (partitioned) into component pieces, and then each bit is doubled before being put back together again:

- double 86 = double 80 + double 6 = 160 + 12 = 172
- double 168 = double 100 + double 60 + double 8 = 200 + 120 + 16 = 336
- double 249 = double 200 + double 40 + double 9 = 400 + 80 + 18 = 498

The last example could also be tackled like this:

- double 249 = double 250 − double 1 = 500 − 2 = 498

We saw earlier that multiplying is the same as adding. A simple example is that 4 × 6 is the same as 4 'lots of' 6, which is 6 + 6 + 6 + 6, which equals 24. We can use this same principle − but in reverse − for more complex sums.

For example:

61 + 64 + 60 + 67

With a little guidance (and some simple partitioning) children can be shown that this sum has 4 'lots of' 60 along with some extra 'bits' and could be rewritten as:

60 + 60 + 60 + 60 + 1 + 4 + 0 + 7

Or as:

4 × 60 + (1 + 4 + 0 + 7)

which equals

$$240 + 12$$

which equals

$$252$$

Another example:

$$80 + 81 + 84 + 89 + 82$$

is the same as

$$5 \times 80 + (0 + 1 + 4 + 9 + 2)$$

which equals

$$400 + \quad 16$$

which equals

$$416$$

Multiplying by 10 is generally considered easier than multiplying by 5. In each of the sums below we could multiply by 10 and then halve the result.

$18 \times 5 = ?$ **Answers:**	$18 \times 10 = 180$	Halve it to get 90
$12 \times 5 = ?$	$12 \times 10 = 120$	Halve it to get 60
$32 \times 5 = ?$	$32 \times 10 = 320$	Halve it to get 160

Multiplying by 50 can be calculated in a very similar fashion: multiply by 100 then halve the result.

For example:

$24 \times 50 = ?$ $24 \times 100 = 2400$ then halve it to get 1200

$16 \times 50 = ?$ $16 \times 100 = 1600$ then halve it to get 800

$49 \times 50 = ?$ $49 \times 100 = 4900$ then halve it to get 2450

Another common strategy to make multiplying more manageable is this: **halve** an even number in the problem given, do this simpler multiplication and then **double** the result.

For example:

6×13	is the same as **double (3 × 13)**	**Answer:** 78
12×7	is the same as **double (6 × 7)**	**Answer:** 84
14×11	is the same as **double (7 × 11)**	**Answer:** 154
9×24	is the same as **double (9 × 12)**	**Answer:** 216

The 16-times table can be worked out by doubling the 8-times table.

For example:

                                           Answer:

16×7 is the same as **double (8 × 7)**   **double** $56 = 112$

Similarly the 24-times table (!) can be worked out by **doubling** the 12-times table (which is itself the same as **double** the 6-times table).

And the 36-times table is of course the same as double the 18-times table, which is the same as double the 9-times table!

One way to multiply by 25 in your head is to multiply by 100 and then divide the answer by 4.

For example:

14×25	=	$14 \times 100 \div 4$	=	$1400 \div 4$	=	350
9×25	=	$9 \times 100 \div 4$	=	$900 \div 4$	=	225
82×25	=	$82 \times 100 \div 4$	=	$8200 \div 4$	=	2050

Need to multiply by 15 in your head? Then you could start by multiplying by 10, then halve that result, and then add the 2 parts together.

For example: 15×18 could be calculated like this:

10 × 18 = 180,
5 × 18 = 90 (as it is half of 180),

so

15 × 18 = 180 + 90 = 270

And 15 × 32 could be calculated like this:

10 × 32 = 320,
5 × 32 = 160 (as it is half of 320),

so

15 × 32 = 320 + 160 = 480

Now let's look at a new concept your children may be introduced to around now, before we move on to multiplying with decimal numbers and negative numbers.

Common multiples and lowest common multiples

Common multiples often result in common mistakes! There is a lot of confusion surrounding multiples, factors, highest common factors, lowest common multiples... What does it all mean?

In Year 6 (ages 10–11), some children will be introduced to the concept of '**common multiples**' and '**common factors**', leading on to the idea of lowest common multiples (LCMs) and highest common factors (HCFs) in the early years of secondary school.

Firstly, what are **multiples**? In simple terms, they are just your times tables facts. So the multiples of 6 are: 6, 12 18, 24, 30, 36, 42, 48...

There are an infinite number of multiples, because in this example you would just keep adding 6 to find the next one, and the next – the sequence never ends.

The tenth multiple of 6 would be the tenth one in the list above (or 10 × 6), which is 60. The eighteenth multiple of 6 would be the eighteenth one in the list above (or 18 × 6), which is 108.

Meanwhile, the multiples for 8 are: 8, 16, 24, 32, 40, 48, 56, 64 and so on. So far, so good!

Second, what are **common multiples**? These are the multiples that two numbers have in common. An example will show you what I mean.

We have already seen that the multiples of 6 are:

> 6, 12 18, **24**, 30, 36, 42, **48**…

and that the multiples of 8 are:

> 8, 16, **24**, 32, 40, **48**, 56, 64…

We now look to see if any of the multiples are the same.

And yes… 24 and 48 appear in both lists and so are common multiples of 6 and 8.

There are more common multiples, of course, and if we extended the lists of multiples we would see them.

Finally – what are **lowest common multiples**? Having got this far, we know what the common multiples are for the two numbers, so now we can find the smallest or *lowest* one. For 6 and 8, the lowest common multiple would be 24.

To work through another example:

- Beth is asked to find the lowest common multiple for 3 and 5.
- She starts by listing some multiples of 3 and 5.
- Multiples of 3 are: 3, 6, 9, 12, **15**, 18, 21, 24, 27, 30, 33…
- Multiples of 5 are: 5, 10, **15**, 20, 25, 30…
- Beth now looks to see whether there are any common multiples, and if so which is the smallest or lowest one.
- She sees that they do have some multiples in common and that 15 is the lowest one. **Answer:** 15

Common factors and highest common factors are equally straightforward and explained in the next chapter: Division (see page 251).

Multiplying with decimal numbers

Decimal Multiplication will be introduced to most but not all primary pupils. Children need to have a very secure grasp of the written methods of multiplication with whole numbers before they can be successfully extended to multiplying with decimals.

So how do we extend the written methods to multiplying with decimals? Surprisingly easily, at least when we are multiplying a decimal number by a single digit whole number, which is all that would generally be expected of primary pupils.

The same methods introduced earlier in 'Moving Forward' (see page 171) and at the start of this section apply here too. We just have to be very careful with the placing of the decimal point. For example:

3.92 × 4

Using the box or grid method:

×	4
3.00	3.00 × 4 = 12.00
0.90	0.90 × 4 = 3.60
0.02	0.02 × 4 = 0.08

Add up the component parts (making very sure the decimal points are in line):

```
12.00
 3.60
 0.08
─────
```
Answer: 15.68

Or using an expanded method:

$$
\begin{array}{r}
3.92 \\
\times \quad 4 \\
\hline
\end{array}
$$

3.00 × 4	12.00
0.90 × 4	3.60
0.02 × 4	0.08

15.68

Or using the standard ('our traditional') method, when multiplying by a single digit number:

$$
\begin{array}{r}
3.92 \\
\times \quad 4 \\
\hline
1\overset{3}{5}.68
\end{array}
$$

where the dialogue would be something like this:

- First put the decimal point in the answer exactly beneath the one in the sum.
- 4 × 2 is 8, so write down the 8.
- 4 × 9 is 36. Carry the 3 and write down the 6.
- 4 × 3 is 12. Add the carried 3 to make 15.

The **absolutely crucial** thing here is that the decimal point in the answer is lined up **exactly underneath** the one in the sum.

Moving into the early years of secondary school, children will be asked to multiply one decimal number by another, something like 3.4 × 2.1.

The methods we've seen above don't work quite so straight-forwardly in this kind of calculation.

Imagine:

$$
\begin{array}{r}
3.4 \\
\times \quad 2.1 \\
\hline
\end{array}
$$

Where would the decimal point go in the answer? And do you still need to put a 0 on the top line, even though you are no longer multiplying with **tens**?

To find the answer, traditional long multiplication does still work, but it needs a little more consideration when dealing with decimals. We'll come back to this below.

Personally, I think Napier's Bones works delightfully well for decimal multiplication:

For 3.4 × 2.1

The grid is drawn as shown earlier:

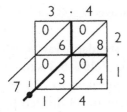

Answer: 7.14

The trick to getting the decimal point in the right place in the answer is simple – when we know how! Put one finger on the decimal point in the number going across the top of the grid and another on the decimal point from the number down the right side of the grid. Run one finger down the vertical line of the grid and the other along the horizontal line, as shown. Where your two fingers meet is the point at which you follow the diagonal line down to the answer. This is where you position the decimal point (see above).

This method works for all decimal multiplication. It is also a method that most children can eventually master – unlike the more standard procedure outlined later.

7.25 × 6.3   0.8 × 3.125

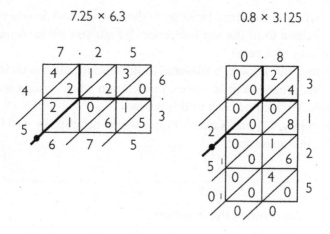

Answer: 45.675   **Answer:** 2.5000 or 2.5

Returning to long multiplication, the standard way your children may be told to tackle something like 7.25 × 6.3 is to rely on them having understood that 7.25 is equivalent to 725 divided by 100 and likewise 6.3 being equivalent to 63 divided by 10. (In reality, this is not always the case – and this can be quite a stumbling block.)

So in theory we have:

$$7.25 \times 6.3 = \frac{725 \times 63}{100 \quad 10} = \frac{725 \times 63}{100 \times 10} = \frac{725 \times 63}{1000}$$

The 725 × 63 are then multiplied in the traditional way:

```
        725
    ×    63
      ¹ ³
    43500
      ¹
     2175
    ─────
    45675
```

And then finally 45675 is divided by 1000, giving the answer 45.675 (see also page 240).

Here is a more challenging example:

6.17 × 4.13

Once again this relies on a child having understood that 6.17 is equivalent to 617 ÷ 100 and likewise 4.13 being equivalent to 413 ÷ 100.

$$6.17 \times 4.13 = \frac{617 \times 413}{100 \quad 100} = \frac{617 \times 413}{100 \times 100} = \frac{617 \times 413}{10\,000}$$

The 617 × 413 are then multiplied in the traditional way:

$$
\begin{array}{r}
617 \\
\times \quad 413 \\
\hline
24\overset{2}{6}800 \\
6170 \\
18\overset{2}{5}1 \\
\hline
254821 \\
\hline
111
\end{array}
$$

Finally 254 821 is divided by 10 000, giving the answer 25.4821.

There is a lot here for children to process. So, using shortcuts can be an invaluable starting point. Here is one such shortcut or 'trick of the trade':

- For 6.17 × 4.13, start by ignoring the decimal point and do 617 × 413 (as above). Now count the total number of digits that come *after* (to the right of) the decimal points in the sum. In this case 6.**17** × 4.**13** has four in total.

This is the number of digits that must come *after* the decimal point in the answer.

Now $617 \times 413 = 254821$

So $6.17 \times 4.13 = 25.\mathbf{4821}$

So for 7.25×6.3 we could simply do 725×63 (which, from the example above, we know is 45675) and then 'put back' the decimal point. In total there are 3 digits *after* the decimal points in the sum, so there must be 3 digits *after* the decimal point in the answer: $7.25 \times 6.3 = 45.675$

This 'trick' will always work, but you must count **all** the digits after the decimal point – even the noughts.

For 0.05×0.125, the standard way would look like:

$$0.05 \times 0.125 = 5 \qquad \frac{5}{100} \times \frac{125}{1000} = \frac{5 \times 125}{100\,000} = \frac{625}{100\,000} = 0.00625$$

Or using the shortcut we would do 5×125 (which is 625) and then count the number of digits *after* the decimal point in the sum:

For $0.\mathbf{05} \times 0.\mathbf{125}$ there are 5 in total.

Therefore there must be 5 digits *after* the decimal point in the answer.

So $0.05 \times 0.125 = 0.\mathbf{00625}$

(As there are not enough digits initially – 625 being only a 3-digit number – we have to 'fill' the necessary places with noughts. There's much more about this in Chapter 5: Division. For now, though, it is sufficient just to notice that the 'fill-in' noughts come *before* the digits, so in this case before the digits 625.)

There will be some who disagree strongly with the idea of 'counting digits' and consider it meaningless nonsense. In principle, I agree with the philosophy of always teaching the full and proper methods so that understanding is the bedrock of learning.

However, at times I do think there can be so much going on at once that the end point becomes obscured and too difficult for many children to reach. In this case shortcuts can actually help children achieve a fuller understanding later on. In fact, once children have mastered the method and can actually multiply two decimal numbers together, unpicking why counting digits 'works' can actually *help* their understanding.

Multiplying with negative numbers

Most children will probably be introduced to this when they reach secondary school.

- Remember: negative numbers can be written with or without brackets. So, for example, 'negative 3' can be written as (-3) or as -3.

Let's look at:

$$4 \times (-2)$$

Well $4 \times (-2)$ means '4 lots of (-2)' and can be written as $(-2) + (-2) + (-2) + (-2)$.

From our knowledge of adding with negative numbers (see also page 102) we know that

$$(-2) + (-2) + (-2) + (-2) \text{ equals } -8 \,.$$

So,

$$4 \times (-2) = -8$$

Likewise for:

$$5 \times -3:$$
$$5 \times -3 \quad = \quad -3 + -3 + -3 + -3 + -3 \quad = \quad -15$$

An alternative approach is to think of –3 as debt of £3, so 5 × –3 is like owing £3 five times over... resulting in a total debt of £15 (that is –15).

- What about (–10) × 6?

As we have seen earlier, multiplications can be done in any order (the commutative rule) so

(–10) × 6 is exactly the same as 6 × (–10)

And

$$6 \times (-10) = -10 + -10 + -10 + -10 + -10 + -10 = -60$$

So what about (–3) × (–7), where we are multiplying *two* negative numbers?

The answer will be positive and so (–3) × (–7) = 21.

But why is the answer positive, when we are multiplying with two negative numbers? This one often catches children out, but it is always the case.

Multiply two negative numbers together and the answer will **always** be positive.

I will try to offer an explanation.

First, let's think about everyday language and double-negatives.

- 'Year 7, you must **not stop** working!' in other words 'Year 7, continue working please.'
- 'Oliver please! Do **not stop** eating!' in other words 'Oliver, eat up please.'

We are familiar with double-negatives and the fact that the outcome or product of two negative things is positive. Well this holds true in mathematics as well as in language.

It's not always easy to grasp this concept at this stage and children may just need to **learn** the rules:

- A positive number multiplied by a positive number gives a **positive** answer.
- A positive number multiplied by a negative number gives a **negative** answer.
- A negative number multiplied by a positive number gives a **negative** answer.
- A negative number multiplied by a negative number gives a **positive** answer.

Often these rules are summarised in a table like the one below:

+	×	+	→	+
+	×	−	→	−
−	×	+	→	−
−	×	−	→	+

So if the signs are the *same* (both positive or both negative), the answer will be positive.

And if the signs are *different* (one positive and one negative), the answer will be negative.

For our children it may be a case of – heads down and just learn it!

Here are some examples:

$$6 \times 3 = 18$$
$$6 \times -3 = -18$$
$$-6 \times 3 = -18$$
$$-6 \times -3 = 18$$
$$4 \times (-5) = -20$$
$$-9 \times 3 = -27$$
$$(-8) \times (-2) = 16$$
$$(-7) \times 4 = -28$$
$$-2 \times -8 = 16$$
$$-3 \times 10 = -30$$

A similar table to the one above also holds true when dividing with negative numbers. Division is explored in detail in the next chapter.

5.
DIVISION

This is the fourth of the 'four operations of number'. **Division** is often seen as the hardest of the four – with some justification. However, just knowing that division is simply the **inverse** (or opposite) of multiplication can help a lot. And having a good grounding in the other three operations will also hold you in good stead.

Division (along with multiplication) tends to be introduced to children a little later than addition and subtraction. As you will see below, children need to be able to add, subtract *and* multiply before they can be taught to divide!

Long division is the 'end game' for many pupils, but it is worth noting that the long division taught today is quite different from the long division many of us will be familiar with from our own school days. It can therefore take some getting used to and 'getting our heads round'. However, as for all the written methods we've seen in previous chapters, this end point is built up gradually in stages.

Children will be taught these different stages over several years and only be moved on from one stage to the next when they are ready. If the new method of long division is unfamiliar to you, you may also like to follow the stages in a step-by-step fashion.

In this chapter you will encounter 'new' terminology such as **'repeated subtraction'** and **'chunking'**. It is all simply explained and I'm sure it will make perfect sense once you have read it. As a very rough rule of thumb, children may divide by using:

- 'repeated subtraction' in Year 2,
- simple 'chunking' in Year 4, and
- more standard methods in Year 6.

UNDERSTANDING THE BASICS
(RECEPTION AND YEARS 1 AND 2, AGES 5–7)

Children are likely to hear and use many different terms that relate to division: share, group, divide, equal groups of, halve, divided by, divided into, divisible by, sharing equally, grouping, repeated subtraction, chunking, division, three (or four or five...) each, divisor, factor, remainder, quotient...

Division tends to be introduced later than addition and subtraction, commonly at the start of Year 2 (around the age of 6). Prior to this they may have had some experience of halving numbers or sharing things into groups.

The rest of this section relates to what most children will begin to experience from Year 2.

The concept of division can be explained in two different ways. As adults we probably use a blend of both strategies without thinking. Young children, discovering division for the first time will have no such awareness. Showing children that division can be 'pictured' in two alternative ways – as outlined below – can promote real understanding.

Sharing equally

Children are usually introduced to the idea of 'sharing' within a 'story' context. For example:

- 'There are 10 sweets in this bag. These 2 children have birthdays today so we are going to share the sweets equally between them. How many sweets will our birthday children have each?'

This can be shown (or *modelled*) as 10 sweets being shared equally, by giving one sweet to each child in turn until there are none left. Once the 'handing out of sweets' is complete the number given to each child is counted. Each child has 5 sweets and this is the answer.

Children can then be introduced to the division symbol ('÷') and shown how to record this sum. In this case, $10 ÷ 2 = 5$.

> An alternative to the division symbol (÷) is '/' (as in ½) or '—' (as in $\frac{1}{2}$)
>
> So the above sum could also be recorded as $^{10}\!/\!_2 = 5$, or $\frac{10}{2} = 5$

Another example:

🔵 There are 15 marbles in this bag. The 3 of you can share them equally.
The 15 marbles can be handed out one at a time until they are all gone. Each child would look to see they have 5 marbles.
Answer: 5 $15 ÷ 3 = 5$

Repeated subtraction (or grouping)

Another way to picture the sum '$10 ÷ 2$' is to keep taking 2 away from 10 until you can't do it any more. Then count up how many times you have taken 2 away.

$$10 - 2 = 8$$
$$8 - 2 = 6$$
$$6 - 2 = 4$$
$$4 - 2 = 2$$
$$2 - 2 = 0$$

Here 2 was taken away (subtracted) 5 times.

Answer: 5 $10 ÷ 2 = 5$

This is the same as saying 'How many twos make 10?' In other words, the children are grouping the 10 into 'lots of' 2.

To start with, children may well draw a 'circle' or 'box' around each group of (or 'lot of') 2 until they can't draw any more. Then they simply count up the number of circled groups to give the answer. This will be easier than doing the actual subtraction sums.

Here is another example of repeated subtraction, for the sum:

$15 \div 3$

Keep taking away 3 until you can't take away any more. Then count up how many times you have taken 3 away.

$15 - 3 = 12$
$12 - 3 = 9$
$9 - 3 = 6$
$6 - 3 = 3$
$3 - 3 = 0$

3 was taken away (subtracted) 5 times.

Answer: 5 $15 \div 3 = 5$

This is the same as saying 'How many threes make 15?' In other words, grouping the 15 into separate 'lots of 3'.

The differences between 'sharing' and 'repeated subtraction' can sometimes seem subtle. A real-life example that demonstrates 'sharing' is when dealing out a pack of cards. Each player is dealt a card from the pack until there are none left. Then each player looks at their 'hand' to see how many cards they have been dealt.

If children can't actually *model* the dealing of the deck, they may record jottings as if they really *were*. Imagine a pack of 52 cards being dealt to 4 friends: Louie, Rico, Alfie and Harry. Children may make a dot or a mark on their paper to show how the cards are being distributed and simply keep counting – and sharing – until they have counted up to 52.

First Louie is dealt a card, then Rico, then Alfie, and then Harry, before Louie is dealt his second card and so on... Jottings might look like this:

Louie	Rico	Alfie	Harry
☐	☐	☐	☐
☐	☐	☐	☐
☐	☐	← At this point 10 cards have been dealt	
		... ☐	☐
☐	☐	☐	☐
☐	☐	☐	☐
☐	☐	☐	☐
☐	☐	☐	☐
☐	☐	☐	☐
☐	☐	☐	☐
☐	☐	☐	☐
☐	☐	☐	☐
☐	☐	☐	☐
☐	☐	☐	☐

Now all 52 cards have been dealt.

Adding up the 'cards' in each column shows that each friend has 13 cards.

So when 52 cards are dealt to 4 people, each person has 13 cards in their 'hand'.

$52 \div 4 = 13$

'Repeated subtraction' is typically seen in a money scenario. Imagine this:

- 'Sophie has been given a mobile phone for her birthday, along with £52 in cash to pay for top-up credits (£1 for each week of the year). Each top-up costs £4. How many times can Sophie top up her phone?'

The calculation is 52 ÷ 4

Each time Sophie tops up, her cash total will reduce by £4. In other words £4 will be repeatedly subtracted until all the money has gone:

$$52 - 4 = 48$$
$$48 - 4 = 44$$
$$44 - 4 = 40$$
$$40 - 4 = 36$$
$$36 - 4 = 32$$
$$32 - 4 = 28$$
$$28 - 4 = 24$$
$$24 - 4 = 20$$
$$20 - 4 = 16$$
$$16 - 4 = 12$$
$$12 - 4 = 8$$
$$8 - 4 = 4$$
$$4 - 4 = 0$$

Four has been subtracted 13 times, therefore 52 ÷ 4 = 13. Sophie can top up her phone 13 times.

Alternatively children may consider how many 'lots of' 4 make 52. And if they know their times tables well they may be able to do this. This is the same as **grouping** 52 into 'lots of' 4. And 13 'lots of' 4 make 52.

You can see straightaway why knowing your times tables is an important stepping stone for successful division.

When children first start dealing with division it is important for them to appreciate that division is **non-commutative** (that is 15 ÷ 3 is *not* the same as 3 ÷ 15). They don't need to know this phrase but they do need to realise from the word go the very important rule, that division CANNOT be done in any order.

To use the example above: $15 \div 3$ is very different from $3 \div 15$ because $15 \div 3 = 5$ whereas $3 \div 15 = 0.2$

One reason why children may find division a little trickier than multiplication is that with multiplication it *doesn't* matter which way round the numbers go:

3×5 is exactly the same as 5×3

Another rule of division that children need to be familiar with is that division is the **inverse** (or opposite) of multiplication and vice versa.

Again, they don't need to know the word but they will, in time, need to appreciate that: division is the opposite of, and so will 'undo', multiplication; just as multiplication is the opposite of, and will 'undo', division.

For example, if a group of four children had 20 sweets and they wanted to share them fairly, they would share or divide 20 by 4.

Each child would then happily have 5 sweets.

Now one of their mums comes along and says '*How many sweets have the 4 of you eaten?*'

One of the children quickly says '*Ah, easy. That's 4 lots of 5 and 4 times 5 is 20.*'

There is more on the inverse rule later in the chapter.

Teachers will sometimes ask children to write **number sentences**. Imagine Daisy had been doing some division sums. She decided she wanted to share 15 marbles between each of her 3 friends, Zoe, Max and Tilly. So she knew she needed to share the marbles equally. Daisy divided and realised each friend would have 5 marbles. '*Good work Daisy!*' says the teacher. '*Can you now record that as a number sentence?*'

Division number sentences use the divide ('$\div$') and equals ('$=$') signs.

So Daisy writes: $15 \div 3 = 5$. And that's it!

Using symbols

Number sentences can also include symbols such as ■ and ▲ to stand for numbers. The task is to work out what number each symbol could be. This is a really useful exercise to reinforce the principle that division is **non-commutative**.

For example:

$20 \div 5 = $ ■   Answer: ■ $= 4$

$9 \div $ ▲ $= 3$ Answer: ▲ $= 3$

⬣ $\div 4 = 2$ Answer: ⬣ $= 8$

⬣ $\div $ ▲ $= 9$ Possible answer: ⬣ $= 18$ ▲ $= 2$

STRUGGLING WITH DIVISION

Why do children (and adults) tend to struggle more with division than multiplication?

This is a really good question but I believe there are tangible reasons for this being so.

First, multiplication is practised much more. Most children in most schools learn their times tables from an early age. But in the past, the link between multiplication and division has often not been stressed – especially at this important formative stage.

Children learning that $7 \times 8 = 56$ should simultaneously be taught that $56 \div 8 = 7$ and that $56 \div 7 = 8$. There are some who say that division is the poor partner of multiplication and that the Monday morning multiplication times tables test should be replaced with a daily division tables test.

Something like this:

$24 \div 6 = ?$

$60 \div 6 = ?$

$18 \div 6 = ?$

$42 \div 6 = ?$

$54 \div 6 = ?$

$30 \div 6 = ?$ ...and so on.

Second, division doesn't always give whole number answers, so sometimes we can stumble into the realm of fractions and decimals when we are least expecting to.

- Multiply 2 whole numbers and the answer will always be a whole number.
- Add or subtract 2 whole numbers and the answer will also always be a whole number.
- Divide 2 whole numbers and the answer might be a whole number or it might not! For example $18 \div 6 = 3$ but $18 \div 8 = 2.25$.

In the early years children avoid fractions and decimals while dividing, by using '**remainders**'. This is like a handy get-out clause to keep things simple and manageable.

For example: $17 \div 5$

Answer: '3 remainder 2'

Third, as we have seen, unlike multiplication, division is **non-commutative** so children need to pay close attention to the order of the numbers. This is especially so with 'wordy' type questions, for example:

'A pencil costs 15p. How many can you buy with 60p?'

$60 \div 15 = 4$

Answer: 4 pencils can be bought with 60p.

Times tables

What it all boils down to is that the key to confident division is to be really, really, really good at your times tables. Know them inside out, upside down and back to front.

Children *will* start learning their times tables at this stage. It is the general aim for children to know their 2-, 5- and 10-times tables by the age of 7 (the end of Year 2).

If children not only know their times tables but also understand them, then division becomes a lot more 'doable'. Knowing that 9 × 5 = 45 *and* understanding that means 9 'lots of' 5 make 45 will lead children naturally to knowing that 45 ÷ 5 = 9 because they will be able to answer the question 'how many fives make 45 – that is, group the 45 into 'lots of' 5.

In exactly the same way they will be able to do 45 ÷ 9 and group the 45 into 'lots of' 9. 45 ÷ 9 = 5

The Number Line again!

The Number Line can also be used for division, in two slightly different ways.

For example:

10 ÷ 2

Children might *start at 10* and jump *back* (to the *left*) in twos until 0 is reached (along similar lines to that of **repeated subtraction** described above). They then count the number of *jumps* to see how many twos make 10.

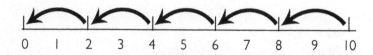

Alternatively, children might *start at 0* and jump up in twos until they reach 10, along the lines of *'how many twos go into 10?'* Again they count the *jumps* to find the answer.

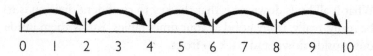

And if there is a remainder?
For example:

$11 \div 4$

Children may *start at 11* and jump *back* (to the *left*) in fours until they can't go any further without passing 0 (along similar lines to that of **repeated subtraction** described above). So they stop at 3 and the number at which they stop *is* the remainder.

Answer: $11 \div 4 = 2$ remainder 3

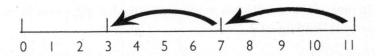

Alternatively, children might start at 0 and jump up in fours. At 8, they stop because they see if they jump another 4 they would go too far and pass 11. The number of jumps are counted and the difference between 8 and 11 is the remainder.

Answer: $11 \div 4 = 2$ remainder 3

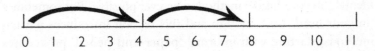

Often when dividing with a bigger number children may be told to use their **Hundred Squares** and count the jumps in a similar fashion to that outlined above. The problem with this is that **Hundred Squares** do not start at 0. So if you are going to use one for division (and they are more compact than a long Number Line) *write a '0'* before the number 1.

Halving

Prior to Year 2, children may be introduced to the idea of halving. By Year 2, children will come to understand that halving is the inverse of doubling. As we saw in the previous chapter, children will soon learn to double all numbers up to at least 20. Now they also need to find the corresponding halves.

For example:

- Find half of 18   Answer: 9
- What is half of 10?   Answer: 5
- Can you find half of 16?   Answer: 8
- Can you share these 20 beads equally,
  between the 2 of you?   Answer: 10

In time, children will learn to appreciate that halving is exactly the same as dividing by 2.

MOVING FORWARD
(YEARS 3 AND 4, AGES 7–9)

We've covered the basics in division above, and all this continues to be used as your children progress, generally using bigger numbers.

It's worth remembering that the aim is for children to use mental ('in your head') methods wherever possible. But sometimes 'in your head' isn't enough and this section introduces written methods, starting with informal **'paper-and-pencil'** procedures and building in stages towards more formal and concise methods.

But first let's have a quick recap from the previous 'Understanding the Basics' section before I introduce a few new concepts.

- Division is **non-commutative** (that is, it cannot be done in any order). For example: 56 ÷ 7 is not the same as 7 ÷ 56.
- Division is the **inverse** of multiplication (and vice versa, so one will 'undo' the other). For example, if 56 ÷ 7 = 8, then 8 × 7 will take you right back to the 56 you started with. This is useful as it allows children to check results and apply their carefully learned times tables to division calculations.
- Following on from the above two rules... children can be encouraged to realise that sometimes they know more than they think they know! Basically, with division and multiplication, if your children know one fact, they actually know four.

For example: if your children *know* 72 ÷ 6 = 12 then whether they realise it or not they also '*know*':

- 72 ÷ 12 = 6 because 6 'lots of' 12 make 72
- 12 × 6 = 72 because multiplication is the inverse of division and
- 6 × 12 = 72 because multiplication is commutative

Repeated subtraction

Sometimes it is easier to subtract than divide. So for 45 ÷ 9, we would repeatedly subtract 9 until we can't do it any more. Then we would add up the number of times we had subtracted 9:

$$45 - 9 = 36$$
$$36 - 9 = 27$$
$$27 - 9 = 18$$
$$18 - 9 = 9$$
$$9 - 9 = 0$$

9 was taken away or subtracted 5 times.

Answer: 5   45 ÷ 9 = 5

Another way to think of this is: '*How many nines make 45?*'

In either case, all we are doing is grouping the 45 into 'lots of' 9. In the first scenario we start at 45 and go backwards until we reach 0. In the second scenario we start at 0 and go forwards until we reach 45.

The Empty Number Line

This can be used to illustrate the above alternative scenarios:

Your children may choose to start at 45 and jump back (to the left) in nines until 0 is reached. They then count the number of jumps (not the numbers) to see how many nines make 45.

0   9   18   27   36   45

Alternatively your children may choose to start at 0 and jump up in nines until they reach 45. Again they count the jumps to find the answer. **Answer: 5**

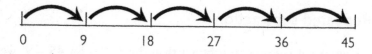

0   9   18   27   36   45

Now follow some new concepts and ideas your children may be introduced to around now.

Dividend, divisor and quotient

I know this sounds like financial jargon that we might prefer to ignore. But these terms actually describe each bit of a division sum. So which bit is the **dividend**, which bit the **divisor** and which bit the **quotient**? Well, in $51 \div 3 = 17$, for example, the 51 is the dividend, the 3 is the divisor and 17 is the quotient.

In practice, pupils rarely use the terms 'dividend' and 'divisor', but as they get older they will need to be familiar with the word 'quotient'.

- The quotient of 18 and 6 is 3 ($18 \div 6 = 3$).
- Similarly the quotient of 12 and 3 is 4 ($12 \div 3 = 4$).

Divisible by...
This is an expression that children *will* need to be familiar with.
For example:

Is 136 divisible by 3?
All that means is:

- 'Can 136 be divided by 3 exactly?' Or 'Can 3 go into 136 an exact number of times with no remainder?'
  Answer: No $136 \div 3 = 45$ remainder 1

Is 261 divisible by 9?
Answer: Yes $261 \div 9 = 29$

Here are some quick tips you might like to know.
A whole number is divisible by:

3	if the sum of its individual digits is divisible by 3. For example, 321 is divisible by 3 because $3 + 2 + 1 = 6$, and 6 is divisible by 3.
6	if the number is even and is also divisible by 3. For example, 162 is divisible by 6 because it is even and also divisible by 3 (because $1 + 6 + 2 = 9$, and 9 is divisible by 3).
8	if the last three digits of the number are divisible by 8. For example, 349816 is divisible by 8 because 816 is divisible by 8 $(816 \div 8 = 102)$.
9	if the sum of the number's digits is divisible by 9. For example, 3285 is divisible by 9 because $3 + 2 + 8 + 5 = 18$, and 18 is divisible by 9.
25	if the last two digits of the number are 00, 25, 50 or 75. For example, 1875 is divisible by 25 because the last two digits are 75.

Dividing by 1

Dividing a number by 1 always leaves a number unchanged.

$6 \div 1 = 6$

0 divided by a number = 0

$0 \div 6 = 0$

This is a rather abstract idea until your children understand that 0 shared by anything is 0. For example, if you have 0 sweets to start with, no matter how many you share that 0 with, there will still be 0 sweets.

THE EXCEPTION TO THE RULE!

$0 \div 0$ does *not* equal 0

Technically $0 \div 0$ is said to be 'indeterminate'

Dividing by 0

Dividing by 0 simply doesn't 'work'. For now it is perfectly accept-able to say something like: '*You can't do it*' or '*It is impossible*'.
    For example:

$6 \div 0 =$ 'impossible'

DIVIDING BY ZERO IS… IMPOSSIBLE!

There may come a point when your children may question why it is impossible (or 'undefined' to be more technically correct). It can be explained by thinking about the way multiplication and division are related:

10 divided by 2 is 5 because	2×5 is 10
10 divided by 0 is 'impossible' because	$0 \times ?$ *is not* 10

This is because there is no value that when multiplied by 0 would give the answer 10. ($0 \times$ any number is always 0.) So when divid-ing by 0 the answer is that… there is no answer!

Understanding the question: the language of division

Even for children who have grasped and understood the important principle that division is **non-commutative**, questions such as those below can cause problems.

- 'A lolly costs 9p. How many can I buy for 54p?'
- 'Jemima and her 3 sisters are given £46 to share equally between them. How much do they each get?'
- 'A chocolate bar costs 25p. How many can I buy with £2?'

Similar problems are also common with **subtraction** (see page 118). As both subtraction and division are non-commutative, more consideration needs to be made to what the question is actually asking and therefore in what order to use the numbers.

Again, spending time ensuring your children have really understood the question is a very important habit to get into. The 'sum' can sometimes take a little unravelling from the 'words'.

Answers to the above questions are:

$54 \div 9 = 6$ 'I can buy 6 lollies.'

$46 \div 4 = 11.5$ 'The 4 sisters (including Jemima) each get £11.50.'

$200 \div 25 = 8$ ($£2 = 200p$) 'I can buy 8 chocolate bars.'

More times tables

Okay, so by the end of Year 2 your children have had some success at mastering their 2-, 5- and 10-times tables. In Year 3 they will also be expected to learn their 3-, 4- and 6-times tables and then know all their tables up to 10×10 by the end of Year 4. Rapid recall of multiplication facts will help enormously when doing the reverse – division.

A lot of encouragement by you – practice, listening to musical renditions of times tables, sticking a big poster-style **Multiplication Square** on the bedroom wall – will all help your children in their ongoing drive to learn their tables off by heart (see also page 164). Being able to rapidly recall multiplication facts makes division much easier.

Dividing by 10 or 100 is an important skill for children to conquer.
For example:

$$40 \div 10 = 4 \qquad 300 \div 100 = 3$$
$$320 \div 10 = 32 \qquad 800 \div 100 = 8$$
$$7000 \div 10 = 700 \qquad 9000 \div 100 = 90$$

Something we were probably all taught at some point was:

- 'Just knock off a nought if you are dividing by 10, and two noughts if you are dividing by 100.' This is a shortcut we probably often use ourselves, so is it okay to tell children this 'rule'?

The problem with this 'rule' is that it works only occasionally, with whole numbers that are themselves a multiple of 10 or 100. It does not work at all with decimal numbers or fractions.
So, $40 \div 10 = 4$ and does *appear* to follow the 'rule':

- 'Just knock off a nought if you are dividing by 10...'

But $47 \div 10$ can't follow this 'rule'. **Answer: 4.7**

Again $300 \div 100 = 3$ and so appears to follow the 'rule':

- '...and two noughts if you are dividing by 100'.

But $30 \div 100$ can't follow this 'rule'. **Answer: 0.3** (See also pages 33 and 43.)
Your children will practise doing a lot of division by 10 and 100. They might discover for themselves when the 'rule' for knocking off noughts works as a short cut. This would be a good 'discovery'.

Related division facts

Using **related facts** can be especially helpful when doing division. This just means using a simpler sum to help with a harder one. For example:

$$21 \div 3 = 7 \qquad \text{so} \qquad 210 \div 3 = 70$$

Why does this work? Because 210 is 10 × as big as 21. As we are still dividing by the same amount, namely 3, the answer must be 10 × as big, which is 70.

Alternatively we could rewrite the sum:

$$210 \div 3 \qquad \text{as} \qquad 10 \times 21 \div 3$$

With '×' and '÷', it doesn't matter which we do first.

So doing the 21 ÷ 3 first, we have: $\qquad 10 \times 7$
Answer: $\qquad\qquad\qquad\qquad\qquad\qquad 70$

Similarly:
$$2100 \div 3 = 700$$
...and $21000 \div 3 = 7000$ ...and so on.

Division and fractions

These are closely connected. In fact you can think of them as close relations.

It is worth remembering that a half of something is the same as sharing it between two. For example: ½ of 8 is the same as 8 ÷ 2.

Likewise a quarter of something is the same as sharing it between four.

For example:

¼ of 12   is the same as   12 ÷ 4
¼ of 2   is the same as   2 ÷ 4

Some more advanced examples:

$\frac{1}{3}$ of 21	is equivalent to sharing between 3:	$21 \div 3$ or $\frac{21}{3}$
$\frac{1}{5}$ of 20	is equivalent to sharing between 5:	$20 \div 5$ or $\frac{20}{5}$
$\frac{1}{8}$ of 24	is the same as sharing between 8:	$24 \div 8$ or $\frac{24}{8}$

And it works the other way round too:

$6 \div 2$	is the same as $\frac{1}{2}$ of 6 or $\frac{6}{2}$
$20 \div 4$	is the same as $\frac{1}{4}$ of 20 or $\frac{20}{4}$
$3 \div 7$	is equivalent to $\frac{1}{7}$ of 3 or $\frac{3}{7}$
$16 \div 3$	is equivalent to $\frac{1}{3}$ of 16 or $\frac{16}{3}$

(See also page 305.)

Finding quarters and eighths – by halving!

Halving and halving again is a quick and easy way to find a quarter of something (and is the same as dividing something into 4). For example:

- One quarter of 160 is 40, because you halve 160 to get 80 and then halve again (half of 80 is 40).

Similarly to find one eighth of something (or to divide something by 8), you halve and halve again, then halve again. For example:

- One eighth of 160 is 20, because you halve 160 to get 80 and halve 80 to get 40 and then halve 40 to get 20.

This technique is a very good one to know and is now very popular in schools.

Partitioning is also used to help with division (see page 74).

Children will need to understand and use the **distributive law** (but not the name).

This means a number can be split or partitioned into two or more 'bits' and then each 'bit' can be divided separately. The partial answers are then **recombined** to give the final answer.

For example:

$96 \div 8$

The 96 is first split or partitioned into 80 and 16. And then each 'bit' is divided by 8. So:

$80 \div 8$ is 10 and $16 \div 8$ is 2.

The 10 and the 2 are then added or recombined to give the final answer of 12. In summary:

$$
\begin{aligned}
96 \div 8 &= (80 + 16) \div 8 \\
&= (80 \div 8) + (16 \div 8) \\
&= 10 + 2 \\
&= 12
\end{aligned}
$$

So why split the number into 80 and 16. Why not 90 and 6? A good question, but we need to remember the whole point of this is to make the division easier to do.

And $80 \div 8$ and $16 \div 8$ are 'easy' divisions because 8 'goes into' them exactly.

Here are some more examples:

$$
\begin{aligned}
87 \div 3 &= (60 + 27) \div 3 \\
&= (60 \div 3) + (27 \div 3) \\
&= 20 + 9 \\
&= 29
\end{aligned}
$$

$$
\begin{aligned}
78 \div 6 &= (60 + 18) \div 6 \\
&= (60 \div 6) + (18 \div 6) \\
&= 10 + 3 \\
&= 13
\end{aligned}
$$

$$
\begin{aligned}
51 \div 3 &= (30 + 21) \div 3 \\
&= (30 \div 3) + (21 \div 3) \\
&= 10 + 7 \\
&= 17
\end{aligned}
$$

Partitioning is certainly helpful for mental division but is also useful for the written methods of division outlined below.

Informal 'pencil-and-paper' methods are usually introduced from Year 4 onwards. Writing things down should simply build on their understanding so far. This will then lead on to **standard written methods**. The written methods are built up in stages, with the aim of helping the children progress to the next stage when they feel confident to do so. The first three stages are outlined in detail below and the final two stages are concluded in the next section: 'To the Top'.

The end point (for some, although not all pupils will reach this stage) is long division, but please note that this method of long division may be quite different from the one you are familiar with from your own school days.

WRITTEN METHODS

For children to use these written methods with ease they must be au fait with the following (all of which were explained in detail earlier in this chapter):

- Understand the terminology. Which bit is the 'dividend', which bit is the 'divisor' and which bit is the 'quotient'? A quick reminder: in $21 \div 3 = 7$, for example, the 21 is the dividend, the 3 is the divisor and 7 is the quotient.
- Understand how to partition. For example, 39 is the same as $30 + 9$.
- Know all times tables (and therefore division facts) up to 10×10.
- Know how to find a remainder. For example, when 43 is divided by 5, the remainder is 3.
- Understand and use multiplication and division as the reverse of each other, that is, the **inverse rule**.
- Know that division can be done by **repeated subtraction**.
- Multiply a 2-digit number by a 1-digit number mentally, such as 12×6.
- Subtract numbers using the **column method** (see page 128).

The stages...

Stage 1: Using partitioning

As we saw above, thoughtful partitioning can help with division. Around Year 4 children might informally record and write down their 'workings'.

For example: $84 \div 7$ might be shown like this:

$$
\begin{array}{ccc}
 & 84 & \\
70 \quad + & & 14 \\
 & & \div\ 7 \\
10 \quad + & & 2 \quad = 12
\end{array}
$$

Or like this:

$$
\begin{aligned}
84 \div 7 &= (70 + 14) \div 7 \\
&= (70 \div 7) + (14 \div 7) \\
&= 10 + 2 \\
&= 12
\end{aligned}
$$

The key to successful division using this method of partitioning is to know the best way to split or partition the number in the first place. Why, for example, did we split the 84 into 70 and 14? Why not into 80 and 4? Well...

We know we are dividing by 7, so we want to split the 84 into bits that we *know* 7 will go into. A good place to start is the multiple of 10 (in other words 10 'lots of' the number we are dividing by). In this case 10 'lots of' 7 is 70. We then see what we are left with, and 14 is the difference between 84 and 70.

The same method can be used even when remainders are involved.

For example: $87 \div 7$ could be recorded like this:

$$
\begin{array}{lll}
 & 87 & \\
70 & + & 17 \\
 & & \div 7 \\
10 & + & 2 \text{ remainder } 3 = 12 \text{ remainder } 3
\end{array}
$$

Or like this:

$$
\begin{aligned}
87 \div 7 &= (70 + 17) \div 7 \\
&= (70 \div 7) + (17 \div 7) \\
&= 10 + 2 \text{ remainder } 3 \\
&= 12 \text{ remainder } 3
\end{aligned}
$$

In the case of $87 \div 7$ we know we are dividing by 7 so we start with 70 (that is 10 'lots of' the number we are dividing by). We then see what we are left with and 17 is the difference between 87 and 70.

Another example: $85 \div 3$ could be written like this:

$$
\begin{array}{lll}
 & 85 & \\
60 & + & 25 \\
 & & \div 3 \\
20 & + & 8 \text{ remainder } 1 = 28 \text{ remainder } 1
\end{array}
$$

Or like this:

$$
\begin{aligned}
85 \div 3 &= (60 + 25) \div 3 \\
&= (60 \div 3) + (25 \div 3) \\
&= 20 + 8 \text{ remainder } 1 \\
&= 28 \text{ remainder } 1
\end{aligned}
$$

Why split the 85 into 60 and 25?

We know we are dividing by 3, so we start with 30 (that is 10 'lots of' the number we are dividing by). But being a whizz with our times tables we then spot we can 'get' another 30 out of 85, so that makes a total of 20 'lots of' 3 (or 60). We then see what we are left with, and 25 is the difference between 85 and 60.

Stage 2: Short division of 2-digit numbers

This can be thought of as a more compact or concise version of recording division using partitioning. This may be introduced at the end of Year 4 or perhaps at the beginning of Year 5.

I include it here as the logic follows on from that set out in Stage 1 – although Stage 3 (below) is probably more predominantly used in Year 4.

For example:

$$51 \div 3$$

Using partitioning – as described above – this could be recorded like this:

$$
\begin{aligned}
51 \div 3 \quad &= (30 + 21) \div \ 3 \\
&= (30 \div \ 3) + (21 \div 3) \\
&= \ 10 + \ 7 \\
&= \ 17
\end{aligned}
$$

Now in Stage 2 we reflect the same logic more concisely.

You may remember (from your own school days) divisions being set out like this:

$$3 \overline{)\ 51} \qquad \text{meaning '51 divided by 3'}$$

(Teachers sometimes refer to $\overline{}$ as the 'bus shelter' symbol.)

Well, this symbol is still seen today to show which number is being divided by what. But now the 'short division' method would *first* look like this:

$$\frac{10 + 7}{3 \overline{)\ 30 + 21}}$$

Answer: 17

You can clearly see the partitioning of 51 into 30 and 21. Each 'bit' is then divided by 3 and the partial answers (10 + 7) are written above the line before being added or recombined to give the final answer of 17.

In time this may be shortened to the more conventional short division and look like this:

$$\frac{1\ 7}{3\overline{)5\ ^2 1}}$$

You may remember doing something similar to this version. Your internal chatter might have been along the lines of:

- '3 into 5 goes once remainder 2. Write the 1 above the line and carry the 2 and put it next to the 1. 3 into 21 goes 7 times. So I write 7 above the line. The answer is 17.'

Simple, yes? Only today's children aren't supposed to say that. Strictly speaking this is how Diva would do this division:

- 'How many threes go into 50 so that the answer is a multiple of 10? Well, 10 threes are 30. So I write 1 above the line representing the 10. The difference between 30 and 50 is 20 so I carry this 20 across (shown by a 2 carry digit) and add it to the 1. So this leaves 21 remaining. And 21 ÷ 3 is 7. So I write 7 above the line. The answer is 17.'

This is more technically correct and mirrors the previous partitioning process beautifully. Nevertheless, once children have reached this stage of doing 'short division' (and have a thorough understanding of place value), doing it 'our way' is a very appropriate aim.

Here is another example: 95 ÷ 7, this time involving a remainder:

$$\frac{10 +\ \ 3 \text{ remainder } 4}{7\overline{)70 + 25}}$$

Answer: 13 remainder 4

This could eventually be shortened to:

$$1 \quad 3 \text{ remainder } 4$$
$$7 \overline{)\ 9\ {}^2 5}$$

Once again, your internal chatter might have been along the lines of:

- '7 into 9 goes 1 remainder 2. Write the 1 above the line and carry the 2 and put it next to the 5. And 7 into 25 goes 3 times, remainder 4.'

Of course Diva will be saying:

- 'How many sevens go into 90 so that the answer is a multiple of 10? Well, 10 sevens are 70. So I write 1 above the line representing the 10. Carry across the difference of 20 (shown with a 2 carry digit) and add it to the 5. And 25 ÷ 7 is 3 with 4 remaining.'

Similarly for 97 ÷ 4:

$$20 + \quad 4 \text{ remainder } 1$$
$$4 \overline{)\ 80 + 17}$$

Answer: 24 remainder 1

This may be shortened to:

$$2 \quad 4 \text{ remainder } 1$$
$$4 \overline{)\ 9^1 7}$$

Your internal chatter might have been along the lines of:

- '4 into 9 goes 2 times remainder one. Write the 2 above the line and carry the 1 and put it next to the 7. Then 4 into 17 goes 4 times, remainder 1. So I write 4 remainder 1 above the line.'

This is how Diva would do this division:

- 'How many fours go into 90 so that the answer is the biggest multiple of 10? It's 20 (because 20 lots of 4 is 80.) So I write 2 above the line in the **tens** column representing the 20. This leaves 10 to carry over and add to the 7. And 17 divided by 4 is 4, with 1 remaining. So I write 4 remainder 1 above the line.'

Another example for $84 \div 3$:

$$
\begin{array}{r}
20 + 8 \\
\hline
3 \,\overline{\,|\, 60 + 24}
\end{array}
$$

or as:

$$
\begin{array}{r}
2\ \ 8 \\
\hline
3 \,\overline{\,|\, 8\,{}^{2}4}
\end{array}
$$

Stage 3: Using chunking!

This may be very new to you. I certainly didn't 'chunk' at school. It's actually very simple and straightforward and is the favoured method for many Year 4 pupils. It simply means: taking away 'chunks' from the number until there is nothing left (along the lines of **repeated subtraction**).

Initially children might take away lots of little 'chunks', but with experience they should be aiming to take away the biggest 'chunks' possible. With this method *no* partitioning is involved. For example:

$97 \div 3$

Now you couldn't really expect a child (or anyone else for that matter) to keep subtracting 3 (along the lines of repeated subtraction outlined in 'Understanding the Basics'). Instead we look to subtract bigger 'chunks'. To start with we might subtract 'chunks' of 30 (which is 10 'lots of' 3, that is 10×3):

```
      97
 −    30    10 × 3      (10 'lots of' 3)
      ──
      67
 −    30    10 × 3      (10 'lots of' 3)
      ──
      37
 −    30    10 × 3      (10 'lots of' 3)
      ──
       7
```

At this point we can no longer subtract 30 but we can take away another 'chunk' of 6 (that is 2×3).

```
 −     6    2 × 3       (2 'lots of' 3)
      ─     ─
       1
```

At this point we can no longer take away any more 'chunks' so the '1' is left as the remainder.
To calculate the answer, add up how many 'lots of' 3 have been subtracted, and write down any remainder.
In this case that is 10 + 10 + 10 + 2 + remainder 1.

Answer: 32 remainder 1

Although perfectly acceptable, this is a bit long winded and not very efficient. As soon as possible, children can be encouraged to reduce the number of steps and so the number of subtractions, by using bigger 'chunks'.

The above example could also look like this:

```
   97
-  60    20 × 3      (20 'lots of' 3)
   ──
   37
-  30    10 × 3      (10 'lots of' 3)
   ──
    7
-   6     2 × 3      (2 'lots of' 3)
   ──     ─
    1
```

Answer: 32 remainder 1

A more efficient recording using 'chunking' may then follow:

```
   97
-  90    (30 'lots of' 3)
   ──
    7
-   6    (2 'lots of' 3)
   ──     ─
    1
```

Answer: 32 remainder 1

As we saw above, the 'bus shelter' symbol '⌐‾‾‾‾' is still in use:

7 ⌐ 84   meaning '84 divided by 7'.

This symbol may be used whilst 'chunking' – it simply shows which number is being divided by what. The method remains exactly the same, as above.

Here are some more examples using 'chunking'...

● 84 ÷ 7:

```
7 | 84
  - 70        10 × 7        (10 'lots of' 7)
  ____
   14
  - 14         2 × 7        (2 'lots of' 7)
  ____
    0
```

Answer: 12

● 168 ÷ 6:

```
6 | 168
  - 60        10 × 6        (10 'lots of' 6)
  ____
   108
  - 60        10 × 6        (10 'lots of' 6)
  ____
    48
  - 48         8 × 6        (8 'lots of' 6)
  ____
     0
```

Answer: 28

● 97 ÷ 4:

```
4 | 97
  - 80        20 × 4
  ____
   17
  - 16         4 × 4
  ____
    1
  ____
    0
```

Answer: 24 remainder 1

🎲 95 ÷ 7:

```
   ____
7 │ 95
  − 70        10 × 7
   ____
    25
  − 21        3 × 7
   ____        −
     4
```

Answer: 13 remainder 4

As discussed elsewhere, it is essential that the **units** line up under **units**, **tens** under **tens**, **hundreds** under **hundreds** and so on.

TO THE TOP
(YEARS 5 AND 6 PLUS, AGES 9–12 PLUS)

As your children continue on their mathematical journey, all the rules and methods for division we've covered in the sections above will continue to be in daily use. Expect bigger and smaller numbers and more complex questions, but the same principles are at work.

Both mental and written methods for division continue to be taught and practised alongside one another. And as we saw in the previous section, written methods are built up gradually in stages. The first 3 stages were introduced in the last section. In this section, we explain the final 2 stages in detail, familiarise ourselves with some more terms and methods, and make the move into decimals and negative (minus) numbers.

At the very beginning of this chapter we saw that division can be understood in two different ways:

• Sharing or
• Repeated subtraction (grouping)

This continues to be the case but now children will be shown that sometimes it is quicker, easier and more efficient to use one way sometimes and the other way at other times. For example, **sharing** is better for dividing by *small numbers*, and repeated subtraction (**grouping**) is better for dividing by *larger numbers*.

For example:

- 24 ÷ 3 would lend itself to sharing: 24 shared equally between 3 is 8.
- 84 ÷ 21 would lend itself to repeated subtraction or grouping:

$$84 - 21 = 63$$
$$63 - 21 = 42$$
$$42 - 21 = 21$$
$$21 - 21 = 0$$

Answer: 4

And as mentioned in the previous section: 'Moving Forward', being able to recall multiplication facts rapidly makes division much easier.

The idea is for all the times tables (up to 10×10) to be known by heart by the end of Year 4. In reality this is not always the case. So more practice, familiarity and just plain old rote-learning will continue throughout Years 5, 6, 7 and beyond.

Dividing makes it smaller...

...when a (positive) number is divided by a number *bigger* than 1.

For example:

$$3 \div 2 = 1.5$$

but,

Dividing makes it bigger...

...when a (positive) number is divided by a positive number *smaller* than 1.

For example:

$$3 \div 0.5 = 6$$

As we've seen in previous sections, division is the **inverse** (reverse) of multiplication. Children should use this fact to check their answers rigorously.

For example, Orla thinks that $72 \div 8 = 9$. To check that she is right, she quickly works out 9×8. Orla is, of course, a whizz at her times tables because she has been practising them religiously since Year 2. And 9×8 is 72 – so job done!

Dividing by 10 (or 100...) is a skill that will have started some time ago but is now extended to include higher powers of 10, such as 1000, and dividing with decimal numbers (see also pages 33 and 43).

For example:

$70 \div 10 = 7$	$900 \div 100 = 9$	$7000 \div 1000 = 7$
$32 \div 10 = 3.2$	$560 \div 100 = 5.6$	$4900 \div 1000 = 4.9$
$5.2 \div 10 = 0.52$	$98 \div 100 = 0.98$	$350 \div 1000 = 0.35$

In the previous section, we asked whether it is okay to teach your child:

- 'Just knock off one nought if you are dividing by 10, two noughts if you are dividing by 100, and three noughts if you are dividing by 1000.'

(See also page 224.) The general answer is 'No!'
Look at:

$$560 \div 100$$

We can't simply say '*knock off two noughts*' because there aren't two noughts to knock off! Instead we need to think about our 'place value' again and think of the number as:

H	T	U	.	t	h
5	6	0			

Dividing by 100 means each digit becomes 100 times smaller, and so the answer is:

H	T	U	.	t	h
		5	.	6	0

Another example:

354.8 ÷ 100

H	T	U	.	t	h	th
3	5	4	.	8	0	0

Answer:

H	T	U	.	t	h	th
		3	.	5	4	8

So

354.8 ÷ 100 = 3.548

You may hear people say:

- 'Just move the decimal point one place to the left if dividing by 10, two places if dividing by 100 and three places if dividing by 1000.'

Is *this* okay to say to our children?

The answer is that this is probably what you and I do and we'd get the right answer, but to be technically correct, we should echo what's generally taught in classrooms now.

That is:

- The *digits* move, but the decimal point stays firmly in its place.
- When dividing, the *digits* move to the **right** (one place if dividing by 10, two places if dividing by 100, three if dividing by 1000 and so on).

Again the best approach is probably to practise doing a lot of division of decimal numbers by 10, 100 and 1000, and see if your children *notice* any pattern. If they do, then they might start working out a 'rule' for themselves.

In the following example we are dividing by 10, so each digit slides to the right one place.

32 ÷ 10

H	T	U	.	t
	3	2	.	0

Answer:

H	T	U	.	t
		3	.	2

So

32 ÷ 10 = 3.2

In this next example we are dividing by 1000, so each digit slides to the right three places.

350 ÷ 1000

H	T	U	.	t	h	th
3	5	0	.	0	0	0

Answer:

H	T	U	.	t	h	th
		0	.	3	5	0

So

350 ÷ 1000 = 0.35

Order matters

In the early years of secondary school (but in Year 6 for some) most children will start doing sums following the **BODMAS** rule (see box).

THE 'BODMAS' RULE

This rule is all about the order in which 'operations' are dealt with when tackling a calculation with several 'operations' embedded within it. Operations, in a mathematical context, are simply something we do to numbers. We have already discussed the 'four operations of number', these being: add; subtract; multiply; and divide. But there are other 'operations'.

'**BODMAS**' is simply an acronym to help us remember in which order to do these operations, each letter being the first letter of the operation:

B	Brackets
O	Order
D	Division
M	Multiplication
A	Addition
S	Subtraction

Possibly the only unfamiliar one is **Order**, meaning powers and roots. For example, 4^2 ('4 squared' or '4 to the power 2' or 4×4) and 5^3 (or '5 cubed' or '5 to the power 3' or $5 \times 5 \times 5$) are powers. The square root of 25 ('$\sqrt{5}$') is 5, that is, 5 is the number that when multiplied by itself makes 25.

BODMAS is simply a way of remembering which bit of a sum to do first. The order is **Brackets** first, *then* **Order**, *then* **Division** *and* **Multiplication** followed by **Addition** *and* **Subtraction**. You will notice that **Division** *and* **Multiplication** have equal ranking, as do **Addition** and **Subtraction**.

Simply working left to right when doing a complex sum will

not necessarily give you the correct answer. Let's look at the following sums:

$$3 + 2 \times 4$$

What answer did you get? Many people would give the *incorrect* answer of 20.

Why is 20 the wrong answer? Because **BODMAS** was not used! Let's look at that sum again:

$$3 + 2 \times 4$$

Using the **BODMAS** rule, multiplication comes before addition – so we must multiply first. And 2×4 is 8 so the sum is now:

$$3 + 8$$

Answer: 11

Here is another example:

$$16 + 9 - 2 \times (9 + 3)$$

To get to the correct answer, you need to apply the **BODMAS** rule. So look for **Brackets**, and do that part of the sum first:

$(9 + 3)$ is calculated first giving the answer 12

Now, **and this is an important bit,** $(9 + 3)$ is *replaced* by 12, and the sum is rewritten as:

$$16 + 9 - 2 \times 12$$

A mistake that a lot of children make is to move the 12 to the start of the sum (thinking that because they did this 'bit' first, it must now come first). This can be avoided if children are encouraged to repeatedly rewrite the sum – underneath the previous one – as they deal with each 'bit'.

After the brackets and following the **BODMAS** rule, we now look for **Orders**. There are none in this sum, so next we look for **Division** or **Multiplication**. There is no division, but there is multiplication:

2 × 12 is calculated giving the answer 24

Once again we rewrite the sum replacing 2 × 12 with 24:

16 + 9 – 24

Now we look for **Addition** and **Subtraction**. These two operations have equal power so now we just do them in the order we see them. In this example **Addition** is encountered first so we do 16 + 9 which is 25 and then rewrite the sum:

25 – 24

So finally we do the **Subtraction**:

25 – 24

Answer: 1

Where do children go wrong with **BODMAS**? All over the place sometimes! It can be quite difficult for children to use **BODMAS** with ease because it goes against what they are familiar with, namely working/reading left to right.

The best way to avoid mistakes is for them to record their workings in a similar way to the examples above, always rewriting the sum.

Some more examples:

$$80 - 6 \times 5 \quad = ? \quad \text{(do the } 6 \times 5 \text{ first)}$$
$$80 - 30 \quad = 50$$

$$8 + 5^2 \times 2 \quad = ? \quad \text{(do the } 5^2 \text{ or } 5 \times 5 \text{ first)}$$
$$8 + 25 \times 2 \quad = ? \quad \text{(followed by the } 25 \times 2)$$
$$8 + 50 \quad = 58$$

$$15 - \% \quad = ? \quad \text{(do the \% first)}$$
$$15 - 3 \quad = 12$$

Try:

$$9 + (1 + 5^2 \times 4) = ?$$

What answer did you get? It should have been the same answer as:

$$9 + 101$$

So you can see that it is not a matter of simply 'doing' the sum in the order you see it. And this is the difference between 'scientific' calculators and cheaper, non-scientific calculators. 'Scientific' calculators use the **BODMAS** rule, the others don't and so throw up incorrect answers. So if you've ever wondered why a cheap calculator you got free with a packet of cornflakes gave a different answer from a scientific calculator, now you know. The answer is the **BODMAS** rule – cheap calculators don't know it! If you are going to get your child a calculator, always choose a scientific model.

As we saw in the previous section, children should be encouraged to realise that sometimes they know more than they think they know!

Basically, with division and multiplication, if your child knows one fact, they actually know 4. This continues to be the case even as the sums appear more complex.

For example: if your children *know* 6 ÷ 12 = 0.5 then they also *know*:

6 ÷ 0.5 = 12 and

0.5 × 12 = 6 (because multiplication is the inverse of division) and also

12 × 0.5 = 6

This might be easier to 'see' with a simpler example: if your children *know* 15 ÷ 3 = 5 then they also *know*:

15 ÷ 5 = 3 and

5 × 3 = 15 (because multiplication is the inverse of division) and also

3 × 5 = 15

Easy halving

Children will be familiar with **halving** being the same as dividing by two.

Now the idea is to find quick and easy ways to halve. In the examples below, each number is broken down (**partitioned**) into component pieces, and then each bit is halved before being recombined or put back together again.

- half of 86 = half of 80 + half of 6 = 40 + 3 = 43
- half of 168 = half of 100 + half of 60 + half of 8 = 50 + 30 + 4 = 84
- half of 249 = half of 200 + half of 40 + half of 9 = 100 + 20 + 4.5 = 124.5

With the last example your children may find it easier or quicker to halve 250 and then adjust accordingly, so 250 ÷ 2 = 125. Now we need to adjust by taking away half of 1. That is, 125 − 0.5 = 124.5.

Factors

This is another term your children definitely need to know.

Factors are simply the numbers you multiply together to make another number.

For example:

$$4 \quad \times \quad 5 \quad = \quad 20$$
$$\uparrow \qquad\qquad \uparrow$$
$$\text{factor} \qquad \text{factor}$$

So 4 and 5 are factors of 20.

Similarly 2 and 10 are also factors of 20 because:

$$2 \quad \times \quad 10 \quad = \quad 20$$
$$\uparrow \qquad\qquad \uparrow$$
$$\text{factor} \qquad \text{factor}$$

Factor pairs are pairs of numbers that multiply together to make the quotient. This is easier than it sounds.

For example **factor pairs** for 20 (the quotient) are:

1 and 20	because	$1 \times 20 = 20$
2 and 10	because	$2 \times 10 = 20$
4 and 5	because	$4 \times 5 = 20$

So '20' has three factor pairs: 1 and 20, 2 and 10, and 4 and 5.

This means '20' has six factors in all, which, in numerical order, are:

1, 2, 4, 5, 10 and 20

IT'S GOT TO BE PERFECT...

There are such things as perfect numbers.

6, 28, 496 and 8128 are all perfect.

What makes them so perfect?

Well, for a number to be perfect, the sum of all its factors other than itself, add up to the number itself, and in the eyes of true mathematicians that's just about as perfect as things can get!

So the factors of 6 are 1, 2, 3 and 6. Add up 1, 2 and 3 to make 6.

The factors of 28 are 1, 2, 4, 7, 14 and 28. Add up 1, 2, 4, 7 and 14 to make 28.

Thought you might like to know. Not everything in life is perfect but 6 is!

A prime number is a number that has only two factors: 1 and itself. For example '11' is prime as only two numbers go into '11' exactly: 1 and 11.

On the other hand '12' is not prime because it has more than two factors, namely: 1, 2, 3, 4, 6 and 12 (see also page 59).

Prime factors

Before considering what prime factors are it might be easier to show you what they are not. For example, if we are thinking about factors for 60, we might consider 6 and 10 because 6 × 10 = 60.

But 6 and 10 are *not* prime factors because they are *not* prime numbers – both have factors of their own. A factor pair for 6 includes 2 and 3 (because 2 x 3 = 6). A factor pair for 10 includes 2 and 5 (because 2 x 5 = 10). Now 2, 3 and 5 *are* prime numbers and hence *are* prime factors.

A factor tree shows this much more clearly.

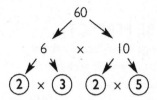

The prime factors are those at the end of the 'branches' of the 'tree' (shown here in bold).

We can write 60 as a **product of its prime factors** by multiplying the prime numbers together:

$$2 \times 3 \times 2 \times 5 = 60$$

We can find 60 as a product of prime factors in different ways – but the end result will always be the same. Here two different factor trees illustrate this:

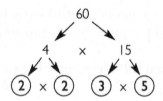

$$2 \times 2 \times 3 \times 5 = 60$$

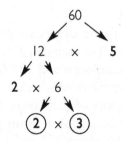

$$2 \times 2 \times 3 \times 5 = 60$$

Common factors and highest common factors

There is a lot of confusion surrounding factors, multiples, highest common factors and lowest common multiples. What does it all mean?

In Year 6 some children will be introduced to the concept of **common factors** and **common multiples**, leading on to the idea of **highest common factors** (HCFs) and **lowest common multiples** (LCMs) in the early years of secondary school.

So what are '**common factors**' and '**highest common factors**'?

First, what are factors? In simple terms… well let me show you:

The factors of 12 are: 1, 2, 3, 4, 6 and 12. They are all the numbers that 'go into' 12 exactly with no remainder. As we saw above, an efficient way to find all the factors of a number is to list 'factor pairs':

$$1 \times 12 = 12$$
$$2 \times 6 = 12$$
$$3 \times 4 = 12$$

There is *always* a finite number of factors for any number.

Secondly, what are common factors? Put simply, these are just the factors that two numbers have in common.

If we wanted to find the common factors for the numbers 36 and 27, for example, we would start by listing all the factors for 36 and then all the factors for 27. Using factor pairs to help:

$$1 \times 36 = 36$$
$$2 \times 18 = 36$$
$$3 \times 12 = 36$$
$$4 \times 9 = 36$$
$$6 \times 6 = 36$$

So the factors of 36, in numerical order, are: **1**, 2, **3**, 4, 6, **9**, 12, 18, 36.

$$1 \times 27 = 27$$
$$3 \times 9 = 27$$

So the factors of 27, in numerical order, are: **1, 3, 9, 27**.

To find the **common factors** for 36 and 27 we look at the lists above to see which numbers appear in both.

The numbers 1, 3, and 9 appear in both lists and so are the *only* common factors for 36 and 27.

Finally what are **highest common factors**?

Well now it's easy. You know what your common factors are for two numbers, so now you just look for the biggest or *highest* one. For 36 and 27, the highest common factor is... (*drum roll*)... **9**.

And that's that!

Carmeena is asked to find the highest common factor for 24 and 32. She starts by listing all the factors of 24 and 32.

- Factors of 24:   1, **2**, 3, **4**, 6, **8**, 12, 24
- Factors of 32:   1, **2**, **4**, **8**, 16, 32

Carmeena now looks to see whether there are any common factors, and if so which is the biggest or highest one. Very quickly she spots that they have several factors in common (shown in **bold**) and that 8 is the highest one. **Answer**: 8

Common multiples and lowest common multiples are equally straightforward (see page 195).

I think one reason why people tend to get confused with these terms is that: **highest common factors** (HCFs) tend to be quite low numbers, whereas **lowest common multiples** (LCMs) tend to be quite high numbers – the opposite of what their titles suggest.

Written methods for division are built up gradually in stages moving on only when children have shown readiness to do so. The first three stages have already been outlined in 'Moving Forward'.

Now we revisit Stage 3, this time with bigger numbers, before looking at the final two stages. Stage 5 is 'long division', the ultimate goal for most pupils, and may be reached in Years 6, 7 or 8 or perhaps not until much later.

There is absolutely no point trying to introduce this stage before children are ready. To be ready children need to have a full grasp of **place value**, the **inverse rule** for division and multiplication, be able to multiply easily and also add and subtract with ease. A big ask! No wonder many people think division is the trickiest of the four operations.

Stage 3 continued...

For an introduction to this, and to the technique of 'chunking' see page 234. For example:

$$173 \div 4$$

Technically speaking we could use **repeated subtraction** to keep taking 4 away until it was no longer possible to do so. Adding up the number of times we'd subtracted 4 would then give us the answer.

But you couldn't reasonably expect a child (or anyone else for that matter) to keep subtracting 4 from a number as big as 173. It would be slow, tedious and prone to mistakes – simply too inefficient. Instead we look to subtract by '**chunking**'. To start with we might subtract 'chunks' of 40 (that is 10 'lots of 4'):

```
    173
 -   40     10 × 4
    ----
    133
 -   40     10 × 4
    ----
     93
 -   40     10 × 4
    ----
     53
 -   40     10 × 4
    ----
     13
```

- At this point we can no longer subtract 40 but we can take away another 'chunk' of 12 (that is 3 lots of 4).

$$
\begin{array}{rl}
- & 12 \quad 3 \times 4 \\
\hline
& 1
\end{array}
$$

- At this point we can no longer take away any more 'chunks' so the '1' is left as the remainder.

Answer: 43 remainder 1

To calculate the answer, add up how many 'lots of' 4 have been subtracted and write down any remainder.

Although perfectly acceptable, this is still a bit long-winded and not very efficient. As soon as possible, children can be encouraged to reduce the number of steps, and therefore the number of subtractions, by using bigger 'chunks'.

The above example could also look like this:

$$
\begin{array}{rl}
& 173 \\
- & 80 \quad 20 \times 4 \\
\hline
& 133 \\
- & 80 \quad 20 \times 4 \\
\hline
& 13 \\
- & 12 \quad 3 \times 4 \\
\hline
& 1
\end{array}
$$

Answer: 43 remainder 1

And also like this:

```
   173
 - 160      40 × 4
   ───
    13
 -  12       3 × 4
   ───       ─
     1
```

Answer: 43 remainder 1

In fact, before starting to calculate a division like 173 ÷ 4, children will first be asked to **approximate** the answer, or work out roughly what the answer will be (see also page 41). To do this they would consider numbers **either** side of 173 that are easy to divide by 4. Both 160 and 200 are easy to divide by 4. Using the related facts that 16 ÷ 4 = 4 and 20 ÷ 4 = 5 then 160 ÷ 4 = 40 and 200 ÷ 4 = 50. As 173 lies somewhere between 160 and 200, then the answer must lie somewhere between 40 and 50. Doing this approximation also helps to 'decide' what the first big 'chunk' could be (namely 160 in this case, which is 40 'lots of' 4).

Here is another example:

197 ÷ 6

Your children may first approximate the answer by knowing that 180 ÷ 6 = 30 and 240 ÷ 6 = 40 (using the related facts that 18 ÷ 6 = 3 and 24 ÷ 6 = 4). From this they would know that their answer must lie somewhere between 30 and 40. Having done this approximation they may choose to subtract the first big 'chunk' of 180 (that is, 30 'lots of' 6). In this example the 'bus shelter' symbol is used – to show which number is being divided by what (see page 236). But the method remains exactly the same.

```
 6 ⌐ 197
  - 180      30 × 6
    ───
     17
  -  12       2 × 6
    ───       ─
      5
```

Answer: 32 remainder 5

And yes! The answer does lie somewhere between 30 and 40.

Stage 4: Short division of 3-digit numbers

Short division of 2-digit numbers *may* be introduced to some children in Year 4 – and therefore the introduction to this is found in Moving Forward (see page 231).

Now bigger numbers are introduced and by the end of Year 5 or beginning of Year 6 most children will start doing short division of a 3-digit number, for example 192 ÷ 6.

In this Stage chunking and repeated subtraction are *not* used. The children will use the technique of partitioning (see also page 74).

Therefore the process of 'working out' 192 ÷ 6, might *first* look like this:

```
    30 +  2
6 | 180 + 12
```

Answer: 32

Here the partitioning is clearly illustrated with 192 first partitioned into 180 and 12 . Then each partitioned 'bit' is divided by 6.

So 6 'goes into' the first bit (180) 30 times and then into the second bit (12), 2 times. The 30 and the 2 are then added (or recombined) to give the answer 32.

This may eventually be shortened to the more conventional layout of short division:

```
    3 2
6 | 1 9 '2
```

Our internal chatter might be along the lines of:

- '6 into 1 won't go; so 6 into 19 goes 3 times remainder one. Write the 3 above the line and carry the 1 and put it next to the 2. Now 6 into 12 goes 2 times.'

As mentioned in the previous section, today's children may say:

- 'How many sixes go into 190 so that the answer is the biggest multiple of 10? 10 sixes are 60, 20 sixes are 120 and 30 sixes are 180. So 30 sixes it is. So I write 3 above the line in the tens column representing the 30. The difference between 180 and 190 is 10 so I carry this 10 across (shown by a 1 carry digit) and add it to the 2. So this leaves 12 remaining. And 12 divided by 6 is 2. So I write 2 above the line. The answer is 32.'

As we saw before, this is the much more technically correct version that accurately reflects the previously practised partitioning process, but still... As long as children have a good understanding of place value and the inverse nature of division and multiplication, doing it 'our way' is an appropriate aim.

Here is another example, with a remainder involved...

The 'working out' for 291 ÷ 8, might *first* look like this:

$$30 + 6 \text{ remainder } 3$$
$$8 \overline{\smash{\big)}\, 240 + 51}$$

Answer: 36 remainder 3

The partitioning of 291 is clearly illustrated with 240 being 30 'lots of' 8 and 51 being the 'bit' left over.

This would eventually be shortened to:

$$3\ 6 \text{ remainder } 3$$
$$8 \overline{\smash{\big)}\, 29\,^5 1}$$

Once again our internal chatter might have been along the lines of:

- '8 into 2 won't go; so 8 into 29 goes 3 times remainder 5. Write the 3 above the line in the tens column and carry the 5 and put it next to the 1. Now 8 into 51 goes 6 times, remainder three.'

Stage 5: Long division

Once short division has been mastered, the next step is long division.

For example:

805 ÷ 35: That is, a 3-digit number divided by a 2-digit number.

For many children, this will be introduced in Year 6.

Now for today's version of long division we go 'back' to using the principle of chunking. Please note that the way long division is taught today is *not* the same as many of us would have been taught.

Using long division, the dialogue for 805 ÷ 35 would go something like this:

- '10 lots of 35 is 350, so 20 lots of 35 would be 700 and 30 lots of 35 would be over 1000, which is too big. So I'll start by taking off a chunk of 700 (20 'lots of' 35) and that will leave me with 105. Now 3 'lots of' 35 is 105 so I'll take this chunk away leaving me with nothing, so no remainder. Now I will add up all the 'lots of' 35 I have subtracted and this will give me the answer. The answer is 23.'

```
35 | 805
  -  700     20 × 35
     ____
      105
  -  105      3 × 35
     ____    ___
        0
```

Answer: 23

It can also be recorded like this:

```
            23
            ‾‾‾‾‾‾
    35 | 805
      − 700
        ‾‾‾‾‾
        105
      − 105
        ‾‾‾‾‾
          0
```

The thought process and dialogue would be exactly the same but the 20 and the 3 are now recorded above the line, instead of down the side. Your children may use either style.

As we saw above, before children start to calculate a division like $805 \div 35$, they should first **approximate** the answer. To do this they would consider that $700 \div 35 = 20$ and $1050 \div 35 = 30$ (using the related facts that $70 \div 35 = 2$ and $105 \div 35 = 3$).

From this they would know that their answer must lie somewhere between 20 and 30. Doing this approximation also helps to 'decide' what the first big 'chunk' could be, namely 20 'lots of' 35 (700, in this case).

Here are more examples...

$204 \div 17$ Approximate answer: between 10 and 20
(because $170 \div 17 = 10$ and $340 \div 17 = 20$)

```
    17 | 204
      − 170      10 × 17
        ‾‾‾‾‾
         34
      −  34       2 × 17
        ‾‾‾‾‾
          0
```
Answer: 12

1344 ÷ 24   Approximate answer: more than 50
(because 2400 ÷ 24 = 100 and 1200 ÷ 24 = 50)

```
24 | 1344
  -  1200     50 × 24
     ‾‾‾‾
      144
  -    96     4 × 24
      ‾‾‾
       48
       48     2 × 24
      ‾‾‾      -
        0
```

Answer: 56

896 ÷ 32   Approximate answer: between 20 and 30
(because 640 ÷ 32 = 20 and 960 ÷ 32 = 30)

```
32 | 896
  -  640     20 × 32
     ‾‾‾
     256
  -  160     5 × 32
     ‾‾‾
      96
      96     3 × 32
     ‾‾‾      -
       0
```

Answer: 28

AND HERE'S HOW WE USED TO DO IT...

You might be thinking this new way of doing division looks rather cumbersome. Why can't we just show them the way we did it?

As we all do a lot less long division these days, can we in fact remember? It is surprisingly difficult. But in case you were trying to remember how you were taught, this is probably how you and I spent many hours tackling such sums as 934 ÷ 26. (Don't worry if you can't follow it. Very few people ever did!)

```
        35  remainder 24
  26 | 934
       78
      ----
      154
   -  130
      ----
       24
```

And we might have said to ourselves:

- 'How many twenty-sixes go into 9? None. So how many twenty-sixes go into 93? It is 3. So I'll write the 3 above the line – in the tens column.
- Now I multiply: 3 sixes are 18, carry the 1 and write down the 8.
- And 3 twos are 6, add it to the carried 1 to make 7. Write down the 7.
- Now I subtract: 93 − 78 to give the answer 15, which I'll write down.
- Now I "bring down" the next digit, which is a 4. So now I no longer have 15 but 154.
- Next step: how many twenty-sixes go into 154? Mmmmm, let me think… 5.
- So I write 5 at the top next to the 3.
- Now to multiply again: 5 × 6 is 30. Carry the 3 and write down the 0.
- And 5 × 2 is 10, add it to the carried 3 to make 13. Write down the 13.

- Now I subtract again: 154 − 130 gives the answer 24, which I'll write down.
- As there are no more digits to "bring down", 24 remains as the remainder.
- There we have it then. Answer: 35 remainder 24.'

I think I know now why this method is no longer the standard one taught! I stress again: please don't worry if you can't follow the above. You don't need to – it's not taught like that any more. Thank goodness!

And now for how 934 ÷ 26 *is* taught today:

```
26 ⌐934
  −  520    20 × 26
     ───
     414
  −  260    10 × 26
     ───
     154
  −  130    5 × 26
     ───    ─
      24
```

Answer: 35 remainder 24

The dialogue might go something like:

- '10 lots of 26 is 260 so I know 20 lots of 26 must be 520. So I'll start by taking off a chunk of 520 (20 lots of 26). Now that leaves me with 414. Now 10 lots of 26 is 260 so I'll take this chunk away, leaving me with 154. Now I can take away 5 lots of 26 which is 130. This will leave 24 as the remainder.'

This can also be recorded like this:

```
          35 remainder 24
26 | 934
   -  520

       4|4
   -  260

       |54
   -  |30

        24
```

Although it looks more like 'our' old method, it isn't. It is simply a slightly more succinct style of the above. The numbers are now recorded above the line, instead of down the side. The logic and dialogue remain exactly the same.

Dividing with decimal numbers

The same methods apply here. We just have to be very careful with the decimal points, which **must** line up under each other.

Children may **not** be expected to handle divisions like these until well into secondary school. I include them here for the sake of completion and to show the same methods are still sound.

For example, 94.5 ÷ 7, using long division with 'chunking', would look like this:

```
7 | 94.5
  -  70.0    10 × 7    Instead of writing 70 we write
                       70.0, so the decimal points can
                       line up underneath each other

     24.5
  -  21.0     3 × 7    Instead of writing 2| we write
                       21.0, so the decimal points can
                       line up underneath each other
```

```
          3.5
   –      3.5    0.5 × 7    one half of 7
         ————    ——
          0.0
```

Answer: 13.5   (that is 10 + 3 + 0.5)

A 'chunk' of 70 was subtracted first, followed by a 'chunk' of 21. This just left 3.5. Spotting that 3.5 is one half (0.5) of 7 was the next step. Adding up the number of times we subtracted 7 gives the final answer of 13.5.

And another example, 106.8 ÷ 8:

```
   8 ⌐ 106.8
   –   80.0    10 × 8    (10 'lots of' 8)
      ————
       26.8
   –   24.0    3 × 8     (3 'lots of' 8)
      ————
        2.8
   –    2.0    0.25 × 8  (one quarter of 8)
      ————
        0.8
   –    0.8    0.1 × 8   (one tenth of 8)
      ————     ————
        0.0
```

Answer: 13.35   (that is 10 + 3 + 0.25 + 0.1)

Dividing with negative numbers

Most children will probably be introduced to this much later on at secondary school.

- Remember: negative numbers can be written with or without

brackets. So for example 'negative 4' can be written as (-4) or as -4.

Let's look at:

$$16 \div (-2)$$

The key to solving this problem is to know that multiplication and division are the **inverse** of each other (that is, division will 'undo' multiplication and vice versa).

- [Quick reminder: $4 \times 3 = 12$
  If we want to 'undo' multiplying by 3, we simply
  divide by 3 and we will be back to where we started
  (that is, 4) $12 \div 3 = 4$]

From our knowledge of multiplying with negative numbers we *know* that: $(-8) \times (-2) = 16$ (See also page 204.)

So, using the inverse rule: $16 \div (-2)$ must take us back to the (-8) we started with.

Therefore: $16 \div (-2) = (-8)$

And if we divide a negative number by a positive number?
For example:

$(-27) \div 3$
Again we know that: $(-9) \times 3 = (-27)$ (See also page 205.)
So, using the inverse rule: $(-27) \div 3$ must take us back to the (-9) we started with.
Therefore: $(-27) \div 3 = (-9)$

- So if a **negative number is divided by a positive number** OR if a **positive number is divided by a negative number**, the answer is ALWAYS **negative**.

So what about –15 ÷ –3, where we are dividing one negative number by another negative number?

The logic remains exactly the same.

We *know* that: $5 \times -3 = -15$ (See also page 203.)

If we now want to 'undo' multiplying by –3 we need to divide by –3.

So, using the inverse rule: $-15 \div -3$ must take us back to the **5** we started with.

Therefore: $-15 \div -3 = \mathbf{5}$

- So if a **negative number is divided by a negative number**, the answer is ALWAYS **positive**.

It's not always easy to grasp this concept at this stage and children may just need to **learn** the rules. It is probably worth noting similar rules apply for multiplication with negative numbers:

- A positive number divided by a positive number gives a **positive** answer.
- A positive number divided by a negative number gives a **negative** answer.
- A negative number divided by a positive number gives a **negative** answer.
- A negative number divided by a negative number gives a **positive** answer.

Often these rules are summarised in a table like the one below:

+	÷	+	→	+
+	÷	−	→	−
−	÷	+	→	−
−	÷	−	→	+

So if the signs are the *same* (both positive or both negative), the answer will be positive. And if the signs are *different* (one positive and one negative), the answer will be negative.

For example:

$$18 \div 3 = 6$$
$$18 \div (-3) = -6$$
$$(-18) \div 3 = -6$$
$$(-18) \div (-3) = 6$$
$$24 \div -8 = -3$$
$$-45 \div 9 = -5$$
$$-36 \div (-6) = 6$$
$$100 \div -10 = -10$$

Please note that **repeated subtraction**, shown earlier in this chapter as a method for division, does **NOT** work when dividing with negative numbers.

So this concludes the big four: Addition; Subtraction; Multiplication; and Division. All will be revisited time and time again in the classroom, to revise, consolidate and affirm children's understanding, each time extending and challenging your children's knowledge – and now with your help too?

6.
NUMBER PATTERNS AND ALGEBRA

Number patterns are all around us – in nature, in art and in the way we use numbers every day. In this chapter we explore how children first encounter them, and how they can help us learn and remember mathematical functions and rules. Perhaps most importantly, we can begin to see how number patterns are the basis of order, harmony and even beauty.

UNDERSTANDING THE BASICS
(RECEPTION AND YEARS 1 AND 2, AGES 5–7)

Children enjoy number patterns from a very early age. Nursery rhymes, songs, stories and counting games all use pattern and rhythm. There are many examples – think of:

- '1, 2, 3, 4, 5, once I caught a fish alive...'
- '5 little ducks went swimming one day...'
- '10 green bottles...'
- '1 man went to mow, went to mow a meadow...'

All these rhymes help children *sequence* numbers, or understand about putting numbers in order, whether counting forward in ones (*such as '1, 2, 3, 4, 5, once I caught a fish alive...'*) *or* backward in ones ('*10 green bottles...*').

So the very first number pattern (or number sequence) children learn is counting on in 'ones'. This can seem so obvious to us that we 'forget' it is a pattern. But a pattern it is, and a very important one. Children will practise counting on in 'ones' in many ways, often starting from different numbers, to help them add and subtract (for example: 8, 9, 10, 11...)

Children like to spot patterns. Often they will notice patterns in all sorts of unusual places: lights on the ceiling, windows on a building, or markings on food packets. You name it, there's probably a pattern on it somewhere – we (grown-ups) just don't 'see' them any more.

Spotting patterns and counting are very closely related when children are very young. Children like to make their own patterns too. They might do this with drawings, but also with different arrangements of objects (like toy cars, Lego bricks, leaves or, infuriatingly, the food on their plate) or with different numbers of jumps or claps (for example, 1 jump, then 2 jumps, then 3 and so on).

Later children learn to count on, and back, in twos. In this way children learn about odd and even numbers, and are introduced to their 2-times tables.

Odd and even numbers are among the first number patterns your children will encounter, and they may well use a Hundred Square like the one below to identify them.

A **Hundred Square** (see also page 18) will probably be used in your children's class. It is a really helpful tool for counting on and spotting patterns.

1	2	3	4	5	6	7	8	9	10
11	12	13	14	15	16	17	18	19	20
21	22	23	24	25	26	27	28	29	30
31	32	33	34	35	36	37	38	39	40
41	42	43	44	45	46	47	48	49	50
51	52	53	54	55	56	57	58	59	60
61	62	63	64	65	66	67	68	69	70
71	72	73	74	75	76	77	78	79	80
81	82	83	84	85	86	87	88	89	90
91	92	93	94	95	96	97	98	99	100

Using a **Hundred Square** we can see that:

- Odd numbers start with the number 1 and then include every other number, like this: 1, 3, 5, 7, 9, 11, 13 and so on.
- They cannot be divided exactly by 2, and they all end with the digits 1, 3, 5, 7 or 9.
- If you do divide an odd number by 2, you will always have 1 left over (that is, a remainder of 1).
- Even numbers start with the number 2 and then include every other number, like this: 2, 4, 6, 8, 10, 12, 14 and so on.
- They are exactly divisible by 2 and all end in a 0, 2, 4, 6 or 8.

By around age 7 (the end of Year 2) children will probably be able to recognise odd and even numbers up to 30, and possibly well beyond.

Counting up in twos, starting at 0, is probably how most children first encounter their **2-times table**. '*2, 4, 6, 8, who do we appreciate?*' is a rhyme lots of children learn to start them on their way.

Children may use a Hundred Square, like the one above, to identify their **2-times table**. Or they may use a **Number Line** or **Multiplication Square**.

A **Number Line** showing the 2-times table may look like this:

This is a number pattern and in time, as children learn all their times tables, they will appreciate that all 'multiples' are number patterns.

A **Multiplication Square** is a big grid showing the first multiples. The grid below shows the first 10 multiples of each of the numbers 1 to 10.

×	1	2	3	4	5	6	7	8	9	10
1	1	2	3	4	5	6	7	8	9	10
2	2	4	6	8	10	12	14	16	18	20
3	3	6	9	12	15	18	21	24	27	30
4	4	8	12	16	20	24	28	32	36	40
5	5	10	15	20	25	30	35	40	45	50
6	6	12	18	24	30	36	42	48	54	60
7	7	14	21	28	35	42	49	56	63	70
8	8	16	24	32	40	48	56	64	72	80
9	9	18	27	36	45	54	63	72	81	90
10	10	20	30	40	50	60	70	80	90	100

For example, the first 10 multiples of 5 are: 5, 10, 15, 20, 25, 30, 35, 40, 45 and 50. These can either be read horizontally along the row or vertically down the column (both highlighted above). Both lines say the same thing, which helps us remember that multiplication is **commutative** (for example: 5 'lots of' 3 is the same as 3 'lots of' 5 or 5 × 3 = 3 × 5).

Another early number sequence that children will practise is counting on – and counting back – in tens: 10, 20, 30, 40 and so on. These are the first 'multiples' of 10 and a multiplication grid like the one above is a good prop for this. But children will also count on in tens starting with *any* number. For example: 23, 33, 43, 53, 63... or 67, 57, 47, 37, 27... In these cases a Hundred Square is more useful.

Look at the **Hundred Square** on page 270. A child can easily be shown how to count on in tens by looking down the column – dropping down 1 row at a time. Starting with the number 23, directly below this number is 33 and this is 10 more than 23. Below 33 is 43 and then 53 and 63 and so on.

Later, children will count on – and back – in hundreds (starting or ending at zero).

An example of counting on: 0, 100, 200, 300, 400...

And an example of counting back: 700, 600, 500, 400, 300, 200, 100, 0.

Once children have understood the concept of number patterns (or number sequences) they will start to **continue** a sequence by **predicting** what comes next. What this means is: they will be shown a number sequence and be asked to describe what they spot or notice. By spotting the pattern, they then see if they can extend or continue the pattern for a few more numbers.

For example:

What do you notice about the following sequence?

2   5   8   11   14   17...

Can you say what the next four numbers will be?

How did you know?

Answer: The children will use their own words to express that each number in the sequence is 3 more than the previous one. So the next 4 numbers are:

20   23   26   29

In this way children are beginning to appreciate that number patterns follow a '**rule**'.

The '**rule**' is the pattern they spot, in this case 'add 3'. They use this 'rule' to **predict** the next 4 numbers, thereby extending or **continuing** the sequence.

Here is another example:

Can you continue this sequence for two more numbers:

28   26   24   22...

Answer: 20, 18. The 'rule' is: take away 2 to get to the next number.

The 'rules' for continuing sequences will be fairly straightforward during the early years of school. Usually the sequences go up (or down) in twos, threes, fives or tens.

As the years go by the 'rules' will steadily become more challenging and complex. (Pupils will still be spotting and extending sequences when they are doing 'A' level maths, only the 'rule' they look for becomes a 'formula' and the next number they look for in the sequence will be the *nth* one – there's a bit more on this below.)

Children will spend a lot of time looking at patterns in their early years of school. Not all of these will be number patterns. But they all help stimulate the part of the brain that makes sense of the world by recognising patterns.

MOVING FORWARD
(YEARS 3 AND 4, AGES 7–9)

Building on from the basics, children will continue to describe and extend number sequences. As we saw earlier, they will count on – or back – in tens and hundreds, but now starting from any number (up to 1000). For example:

Count on in hundreds from 340 to 940. How many hundreds did you count?
You start at 340 then – 440, 540, 640, 740, 840, 940.
Answer: 'We counted 6 hundreds.'

Another example:

Describe this sequence:
854   754   654   554   454…
What comes next?
Answer: *The numbers go down by a 100 each time, and 354 comes next*' (or some similar description in the children's own words).

Number patterns can be made by counting on – or back – in steps of any number you like. Common number patterns likely to be seen in the classroom involve counting on, or back, in steps of 25 or 50. Counting back may now go below 0.

For example:

⬢ Count back in 25s from 300 to −100.
Answer: 300, 275, 250, 225, 200, 175, 150, 125, 100, 75, 50, 25, 0, −25, −50, −75, −100

⬢ Count on in 50s to 1000 and then back again.
Answer: 50, 100, 150, 200, 250, 300, 350, 400, 450, 500, 550, 600, 650, 700, 750, 800, 850, 900, 950, 1000… and then in reverse…1000, 950…

During Years 3 and 4 there will be a big push for pupils to learn their times tables. Some find it difficult, some boring, but the idea is that the hard work involved in learning tables now will pay off later as children confidently use their stored knowledge as part of more complex calculations.

It can help children find tables easier and more interesting to learn if they remember that all multiples are number patterns. For example, the first 10 multiples of 4 are: 4, 8, 12, 16, 20, 24, 28, 32, 36, 40 and this sequence can simply be extended by 'adding 4'.

Children will undoubtedly use patterns on **Hundred Squares** to help them learn their times tables. Often they will be asked to shade in numbers from the start of a times table. Very quickly a pattern emerges. These visual images can be a big help for some children.

A simple pattern highlights the 2-times table:

1	**2**	3	**4**	5	**6**	7	**8**	9	**10**
11	**12**	13	**14**	15	**16**	17	**18**	19	**20**
21	**22**	23	**24**	25	**26**	27	**28**	29	**30**
31	**32**	33	**34**	35	**36**	37	**38**	39	**40**
41	**42**	43	**44**	45	**46**	47	**48**	49	**50**
51	**52**	53	**54**	55	**56**	57	**58**	59	**60**
61	**62**	63	**64**	65	**66**	67	**68**	69	**70**
71	**72**	73	**74**	75	**76**	77	**78**	79	**80**
81	**82**	83	**84**	85	**86**	87	**88**	89	**90**
91	**92**	93	**94**	95	**96**	97	**98**	99	**100**

For the 5-times table, the pattern looks like this:

1	2	3	4	**5**	6	7	8	9	**10**
11	12	13	14	**15**	16	17	18	19	**20**
21	22	23	24	**25**	26	27	28	29	**30**
31	32	33	34	**35**	36	37	38	39	**40**
41	42	43	44	**45**	46	47	48	49	**50**
51	52	53	54	**55**	56	·57	58	59	**60**
61	62	63	64	**65**	66	67	68	69	**70**
71	72	73	74	**75**	76	77	78	79	**80**
81	82	83	84	**85**	86	87	88	89	**90**
91	92	93	94	**95**	96	97	98	99	**100**

And for the 3-times table it looks like this:

1	2	**3**	4	5	**6**	7	8	**9**	10
11	**12**	13	14	**15**	16	17	**18**	19	20
21	22	23	**24**	25	26	**27**	28	29	**30**
31	32	**33**	34	35	**36**	37	38	**39**	40
41	**42**	43	44	**45**	46	47	**48**	49	50
51	52	53	**54**	55	56	**57**	58	59	**60**
61	62	**63**	64	65	**66**	67	68	**69**	70
71	**72**	73	74	**75**	76	77	**78**	79	80
81	82	83	**84**	85	86	**87**	88	89	**90**
91	92	**93**	94	95	**96**	97	98	**99**	100

The pattern for the 4-times table isn't so immediately obvious, but there is one:

1	2	3	**4**	5	6	7	**8**	9	10
11	**12**	13	14	15	**16**	17	18	19	**20**
21	22	23	**24**	25	26	27	**28**	29	30
31	**32**	33	34	35	**36**	37	38	39	**40**
41	42	43	**44**	45	46	47	**48**	49	50
51	**52**	53	54	55	**56**	57	58	59	**60**
61	62	63	**64**	65	66	67	**68**	69	70
71	**72**	73	74	75	**76**	77	78	79	**80**
81	82	83	**84**	85	86	87	**88**	89	90
91	**92**	93	94	95	**96**	97	98	99	**100**

Children can have a go at spotting patterns for any of their times tables. Below is the pattern for the 9-times table:

1	2	3	4	5	6	7	8	**9**	10
11	12	13	14	15	16	17	**18**	19	20
21	22	23	24	25	26	**27**	28	29	30
31	32	33	34	35	**36**	37	38	39	40
41	42	43	44	**45**	46	47	48	49	50
51	52	53	**54**	55	56	57	58	59	60
61	62	**63**	64	65	66	67	68	69	70
71	**72**	73	74	75	76	77	78	79	80
81	82	83	84	85	86	87	88	89	**90**
91	92	93	94	95	96	97	98	**99**	100

Multiplication Squares, which list the times tables in rows and columns, are another very good prop to help children learn their tables. Children can also be shown how to spot patterns and use the symmetry of the square to help them learn their tables.

For example, given a Multiplication Square, children may be asked to find and explain as many patterns as possible. Below is the pattern made when all multiples of '3' are highlighted:

×	1	2	3	4	5	6	7	8	9	10
1	1	2	**3**	4	5	**6**	7	8	**9**	10
2	2	4	**6**	8	10	**12**	14	16	**18**	20
3	**3**	**6**	**9**	**12**	**15**	**18**	**21**	**24**	**27**	**30**
4	4	8	**12**	16	20	**24**	28	32	**36**	40
5	5	10	**15**	20	25	**30**	35	40	**45**	50
6	**6**	**12**	**18**	**24**	**30**	**36**	**42**	**48**	**54**	**60**
7	7	14	**21**	28	35	**42**	49	56	**63**	70
8	8	16	**24**	32	40	**48**	56	64	**72**	80
9	**9**	**18**	**27**	**36**	**45**	**54**	**63**	**72**	**81**	**90**
10	10	20	**30**	40	50	**60**	70	80	**90**	100

And the table below shows the pattern when all multiples of '4' are highlighted:

×	1	2	3	4	5	6	7	8	9	10
1	1	2	3	4	5	6	7	8	9	10
2	2	4	6	8	10	12	14	16	18	20
3	3	6	9	12	15	18	21	24	27	30
4	4	8	12	16	20	24	28	32	36	40
5	5	10	15	20	25	30	35	40	45	50
6	6	12	18	24	30	36	42	48	54	60
7	7	14	21	28	35	42	49	56	63	70
8	8	16	24	32	40	48	56	64	72	80
9	9	18	27	36	45	54	63	72	81	90
10	10	20	30	40	50	60	70	80	90	100

Using the symmetry of the square, a diagonal line as shown below splits the square in half and the times tables are reflected either side of the line (see also page 155).

×	1	2	3	4	5	6	7	8	9	10
1	1	2	3	4	5	6	7	8	9	10
2	2	4	6	8	10	12	14	16	18	20
3	3	6	9	12	15	18	21	24	27	30
4	4	8	12	16	20	24	28	32	36	40
5	5	10	15	20	25	30	35	40	45	50
6	6	12	18	24	30	36	42	48	54	60
7	7	14	21	28	35	42	49	56	63	70
8	8	16	24	32	40	48	56	64	72	80
9	9	18	27	36	45	54	63	72	81	90
10	10	20	30	40	50	60	70	80	90	100

The numbers in the highlighted diagonal line are in fact another number pattern. These numbers (1, 4, 9, 16, 25...) are called **square numbers** and are looked at in more detail on page 284.

Odd and even numbers were introduced in the previous section: 'Understanding the Basics'. Now children may be able to recognise odd and even numbers up to 100 (by the end of Year 3) and up to a 1000 (by the end of Year 4).

As well as identifying odd and even numbers, children will start to discover what happens when odd and even numbers are added or subtracted:

- If two **even** numbers are added together or subtracted from one another, the answer will always be **even**. Examples: 24 + 12 = 36; 14 − 6 = 8.
- This remains so regardless of how many even numbers you add or subtract. Examples: 8 + 32 + 16 + 4 = 60; 32 − 10 − 4 = 18; 6 + 2 − 4 + 10 − 8 = 6.
- If two **odd** numbers are added together or subtracted from one another, the answer will always be **even**. Examples: 15 + 3 = 18; 15 − 3 = 12.

IS 0 ODD OR EVEN?

This is a question often asked by older children.

It is not as easy to answer as it may seem and it often causes much debate.

Some argue it is neither. We talk about an odd number of socks or an even number of dinner guests but 0 just doesn't seem relevant in these kinds of contexts.

On the other hand if we apply the test for 'evenness', 0 appears even. That is, 0 can be divided by 2 with no remainder.

As introduced in 'Understanding the Basics', children will start to describe and extend number sequences. By following a **'rule'** they will start to **predict** which numbers come next in a sequence.

For example:

Describe the following sequence.
37   46   55   64...
What are the next two numbers?
How did you know?
Answer: The children will use their own words to express that
'each number in the sequence is 9 more than the previous one'.
So the next two numbers are: 73, 82.

The **'rule'** is the pattern they spot, in this case 'add 9'. They then
use this 'rule' to **predict** the next two numbers, thereby **extending** or continuing the sequence.

Another example:

Fill in the missing numbers in this sequence.
86   __   78   74   __   66   __
Explain the rule
Answer: 82, 70, 62. The numbers are 4 less each time.

TO THE TOP
(YEARS 5 AND 6 PLUS, AGES 9–12 PLUS)

Everything previously discussed in this chapter will still be in use
as children move forward. Recognising **multiples** up to 10 × 10
and using a Hundred Square and a Multiplication Square to spot
patterns will still be a high priority at this stage.

Children continue to recognise and extend number sequences.
The number sequences may be formed by counting from *any*
number in steps of constant size.

Examples include:

- 'Count in steps of 25, from 800 to 1000 then back.'
- 'Count in steps of 0.1, from 5 to 7 and then back again.'
- 'Count in steps of 0.25 from 0 to 3.'
- 'Count back in steps of 0.5 from 4 to −3.'

Answers:
800, 825, 850, 875, 900, 925, 950, 975, 1000 − and then back − 1000, 975, 950, 925, 900, 875, 850, 825, 800

5, 5.1, 5.2, 5.3, 5.4, 5.5, 5.6, 5.7, 5.8, 5.9, 6.0, 6.1, 6.2, 6.3, 6.4, 6.5, 6.6, 6.7, 6.8, 6.9, 7.0 − and then back − 7.0, 6.9, 6.8, 6.7, 6.6, 6.5, 6.4, 6.3, 6.2, 6.1, 6.0, 5.9, 5.8, 5.7, 5.6, 5.5, 5.4, 5.3, 5.2, 5.1, 5.0

0, 0.25, 0.5, 0.75, 1, 1.25, 1.5, 1.75, 2, 2.25, 2.5, 2.75, 3

4, 3.5, 3, 2.5, 2, 1.5, 1, 0.5, 0, −0.5, −1, −1.5, −2, −2.5, −3

All numbers are either **odd or even** (see page 270). Children will need to recognise odd and even numbers up to at least 1000 and build on the properties 'discovered' earlier. We have previously seen the outcome when we add or subtract odd and even numbers

But what happens if we multiply odd and even numbers?

- The product of two even numbers is even (for example, $4 \times 6 = 24$).
- The product of two odd numbers is odd (for example, $3 \times 5 = 15$).
- The product of one odd and one even number is even (for example, $3 \times 4 = 12$).

Throughout the chapter we have seen how children can describe and extend number sequences. By following a **'rule'** they can **predict** what numbers come next in a sequence.

For example:

Fill in the missing numbers in this sequence.

-30   -27   -24   _   -18   _   -12   _

Explain the rule orally (in words) and in writing.

Answer: -21 -15 -9. The numbers are 3 more each time.

The word '**term**' is often used instead of the word 'number' in a sequence and children *do* need to know this word.

For example:

Write the next two terms in this sequence:

91   72   53   34   _   _

Answer: 15 -4. Each term is 19 less than the previous one.

Sequences do *not* have to go up or down in constant amounts. Look at the following sequence:

1   3   6   10   15   21...

What would the next two terms be?

There is a difference of 2 between 1 and 3, then a difference of 3 between 3 and 6, then a difference of 4 between 6 and 10, then a difference of 5 between 10 and 15, then a difference of 6 between 15 and 21, then...

This can be illustrated like this:

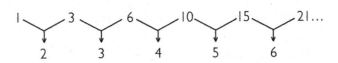

The differences show the pattern. The next difference would be 7, so the next term in the sequence would be 28 (that is, 21 plus 7) and the next term would be 36 (28 plus 8).

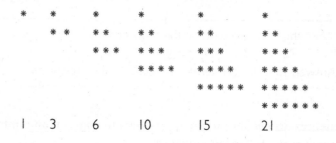

This is a special sequence. The numbers in this pattern are the **'triangular numbers'**.

Triangular numbers are so named because they can form the shapes of triangles. (A little imagination has to be used to see the first dot as a triangle, but its place in the sequence becomes clear.)

```
*       *       *       *         *           *
        * *     * *     * *       * *         * *
                * * *   * * *     * * *       * * *
                        * * * *   * * * *     * * * *
                                  * * * * *   * * * * *
                                              * * * * * *

1       3       6       10        15          21
```

Another very special sequence of numbers is the **Square Numbers**.

Square Numbers are:

  1   4   9   16   25   36   49   64   81   100...

Square numbers are the result of multiplying each number by itself:

  1×1 2×2 3×3 4×4 5×5 6×6 7×7 8×8 9×9 10×10 ...

The next two square numbers are therefore:

  121   and   144
  (11×11)   (12×12)

Square numbers can also be shown pictorially as the numbers of dots can be arranged to make squares. Once again, the first single dot has to be imagined as a square.

```
 *     * *    * * *    * * * *    * * * * *
       * *    * * *    * * * *    * * * * *
              * * *    * * * *    * * * * *
                       * * * *    * * * * *
                                  * * * * *
 I     4      9        16         25
```

Two consecutive **triangular numbers** added together make a **square number**.

For example:

I	+	3	=	4	
3	+	6	=	9	
6	+	10	=	16	
10	+	15	=	25	...and so on.

Square numbers are found in the longest diagonal line of a multiplication grid, and cut the grid exactly in half (see page 279).

So to **square a number** you simply multiply it by itself.

For example:

- 7 squared (or 7^2) is '7×7', which is 49 ($7 \times 7 = 49$).

The opposite or **inverse** of this is to **square root** a number.

For example:

- The square root of 49 is 7 (because $7 \times 7 = 49$).

The square root symbol looks like this:

- '$\sqrt{}$' and so the square root of 49 can be written as $\sqrt{49}$.

Perfect square numbers will give whole number answers when you 'square root' them. The first 12 perfect square numbers are:

I   4   9   16   25   36   49   64   81   100   121   144

The corresponding **square roots** are:

1   2   3   4   5   6   7   8   9   10   11   12

The next perfect square number is 13×13 which is 169. So the square root of 169 is 13.

As with times tables it can really help children if they just '*know*' the first 12 square numbers and their corresponding square roots.

Another very famous number pattern is the **Fibonacci sequence**. This starts:

1   1   2   3   5   8   13   21   34   55   89...

Your children might not necessarily encounter this pattern at primary school, but will most certainly see it at some point in secondary school. It is a beautiful sequence, and one often found in nature. (It is seen in the arrangements of petals on a flower, and in the spirals of a shell, and in a multitude of other natural forms.)

To find the next term in the sequence, you add together the previous two.

1 1 2
$1 + 1 = 2$

1 2 3
$1 + 2 = 3$

2 3 5
$2 + 3 = 5$

3 5 8
$3 + 5 = 8$

5 8 13
$5 + 8 = 13$

8 13 21
$8 + 13 = 21$ ...and so on.

'**Consecutive**' means next to or 'adjacent'. Children will be expected to be familiar with this word in a mathematical context. For example, in the Fibonacci sequence 3, 5 and 8 are consecutive. So too are 8, 13, 21, 34 and 55.

In the sequence of square numbers: 1, 4, 9, 16, 25, 36, 49, 64 and 81... the terms 16 and 25 are consecutive. So too are 49, 64 and 81.

The first four consecutive terms of the sequence: 7, 10, 13, 16, 19, 22, 25, 28, 31, 34... are 7, 10, 13 and 16.

Consecutive numbers are simply our counting numbers.

3 and 4 are consecutive numbers.

7, 8, 9, 10 and 11 are consecutive numbers.

34, 35 and 36 are consecutive numbers.

In their early years at secondary school children may be asked to 'make' all the numbers between 1 and 30 by adding consecutive numbers. I'll show you what I mean:

1	0 + 1
2	
3	1 + 2
4	
5	2 + 3
6	1 + 2 + 3
7	3 + 4
8	
9	4 + 5
10	1 + 2 + 3 + 4
11	5 + 6
12	3 + 4 + 5 ...and so on.

Not all the numbers between 1 and 30 can be made using consecutive numbers. But most can. The numbers that *cannot* be made this way are: 2, 4, 8 and 16 and these numbers are themselves a sequence with a **relationship** (see also page 292).

'**Relationship**' is a word that children will need to be familiar with in a mathematical context. It describes how numbers are connected or related to each other.

A '**formula**' is a more mathematically rigorous version of the description of the **relationship**. A formula is a set of instructions that will always apply, no matter what. It is a rule that always works.

To find a formula you have to look very carefully at the relationship – and this is outlined in detail below.

Pupils in Years 5 and 6 will be introduced to formulae with simple sequences.

For example:

- 'Find the formula for the following sequence: 3, 6, 9, 12, 15...'

We would start by looking for a relationship. In this case the relationship is that each number in the sequence is a multiple of 3.

3	is	1×3
6	is	2×3
9	is	3×3
12	is	4×3 ...and so on.

We would then rewrite this information in a table. The top line of the table shows the position of each term in the sequence. The second line in the table shows the actual terms from the sequence, and the third line highlights the relationship.

Position	1st	2nd	3rd	4th	5th	... nth
Sequence	3	6	9	12	15	...
Relationship	1×3	2×3	3×3	4×3	5×3	...

To find a formula we need to make a connection between the 'position number' and the 'relationship'.

To start with, your children may be able to spot the next numbers in the sequence. The 6th term would be 6×3 (18), the 7th term would be 7×3 (21) and the 8th term would be 8×3 (24). They may also be able to predict terms out of order, such as the 20th term as 20×3 and the 100th term as 100×3.

This is good, but not quite enough for a formula.

NUMBER PATTERNS AND ALGEBRA

For a formula we need to have one 'rule' that works for any and all cases. In other words, we say we want to find the **nth term** of the sequence.

What exactly is the nth term? This is the name we use to refer to 'any' term – the 1st, the 8th, the 27th, or whatever.

It is now just one more mental step to find the nth term.

By spotting patterns and being able to predict terms out of order, your children will have made the connection between the 'position number' and the relationship. In this example the 'position number' reappears as the multiple of 3. For example:

- the 5th term is 5 × 3 and the 7th term is 7 × 3 and the 29th term is 29 × 3.

So what would the nth term be? Simple: n × 3.

Purely for elegance this can then be rewritten as 3 × n and then – if you like – abbreviated to 3n. And this is the **formula** for the sequence.

Using the formula (n × 3) we can now replace 'n' with any position number to find that term of the sequence.

- So if n = 1 the 1st term in the sequence is '1 × 3' which is 3;
- If n = 2 the 2nd term in the sequence is '2 × 3' which is 6;
- If n = 13 the 13th term in the sequence is '13 × 3' which is 39;
- If n = 17 the 17th term in the sequence is '17 × 3' which is 51;
- And if n = 134 the 134th term in the sequence is '134 × 3' which is 402.

Having a formula means we can easily and efficiently 'find' any term in a sequence without having to write out all the preceding terms.
    Here is another example.

- 'Find the formula for the following sequence: 5, 10, 15, 20, 25…'

Again, we would start by looking for a relationship. In this case the relationship is that each number in the sequence is a multiple of 5.

5	is	1×5
10	is	2×5
15	is	3×5
20	is	4×5 ...and so on.

Rewriting this information in a table:

Position	1st	2nd	3rd	4th	5th	...	nth
Sequence	5	10	15	20	25	...	
Relationship	1×5	2×5	3×5	4×5	5×5	...	

Spotting patterns is always a good starting point. The 6th term would be 6×5 (30), the 7th term would be 7×5 (35) and the 8th term would be 8×5 (40).

The 'position number' reappears in the 'relationship' as the multiple of 5.

So what is the nth term?

- Simple: $n \times 5$

This can then be rewritten as $5 \times n$ or as $5n$. And this is the **formula** for the sequence.

And another example.

- 'Find the formula for the following sequence: 6, 11, 16, 21, 26...'

Again, we would attempt to spot a pattern to help us find a relationship. This one needs a little more thought but it should help if we bear the previous example in mind. The numbers all go up by 5 but they are not multiples of 5. In fact it is very similar to the previous formula – with just a little adjustment!

There are two steps needed now and children often work this out with a little guidance.

6   is   1 × 5 and then add 1
11   is   2 × 5 and then add 1
16   is   3 × 5 and then add 1
21   is   4 × 5 and then add 1
...and so on.

Rewriting this information in a table:

Position	1st	2nd	3rd	4th	5th	... nth
Sequence	6	11	16	21	26	...
Relationship	(1×5)+1	(2×5)+1	(3×5)+1	(4×5)+1	(5×5)+1	...

- The 6th term is (6 × 5) + 1
- The 7th term is (7 × 5) + 1
- The 8th term is (8 × 5) + 1

The 'position number' reappears in the 'relationship' as the multiple of 5 and then 1 is added each time.

- The nth term is (n × 5) + 1.

The formula could quite satisfactorily be left like this or it could be rewritten as

$5n + 1.$

And another example:

- 'Find the formula for the following sequence: 1, 4, 9, 16, 25...'

This is a very common sequence and one your children might recognise.

1   is   1 × 1   otherwise known as 1 squared (1^2)
4   is   2 × 2   otherwise known as 2 squared (2^2)
9   is   3 × 3   otherwise known as 3 squared (3^2)

16	is	4 × 4	otherwise known as 4 squared (4^2)
25	is	5 × 5	otherwise known as 5 squared (5^2)

...and so on.

Rewriting this information in a table:

Position	1st	2nd	3rd	4th	5th	...	nth
Sequence	1	4	9	16	25	...	
Relationship	1^2	2^2	3^2	4^2	5^2	...	

- The 6th term is 6^2, the 7th term is 7^2 and the 8th term is 8^2.
- The nth term is n^2

AND NOW FOR A SLIGHTLY MORE CHALLENGING ONE...

Pupils in Year 7 or 8 *may* be able to find the formula for this sequence – but it would be a challenge for many. Looking again at the sequence 2, 4, 8, 16... (see page 287) we can see that the numbers can be rewritten using index notation.

Index notation is a useful way of writing expressions like $2 \times 2 \times 2 \times 2 \times 2$ in a shorter format, namely 2^5 in this case. The small number is called the **index** or **power**.

The relationship for the above sequence is that each number can be written as '2 *to the power of* ...' In other words:

2	is	2	or	2 to the power of 1	or	(2^1)
4	is	2 × 2	or	2 to the power of 2	or	(2^2)
8	is	2 × 2 × 2	or	2 to the power of 3	or	(2^3)
16	is	2 × 2 × 2 × 2	or	2 to the power of 4	or	(2^4)

The next number in this sequence would be:

32 which is
$2 \times 2 \times 2 \times 2 \times 2$ or   2 to the power of 5   or (2^5).

And the next one would be:

64 which is
$2 \times 2 \times 2 \times 2 \times 2 \times 2$ or   2 to the power of 6   or (2^6).

We rewrite all this information in a table. As before, the top line of the table shows the position of each term in the sequence. The second line in the table shows the actual terms from the sequence, and the third line highlights the relationship.

Position	1st	2nd	3rd	4th	5th	...	nth
Sequence	2	4	8	16	32	...	
Relationship	2^1	2^2	2^3	2^4	2^5	...	

Your children may be able to spot or predict that the 6th term in the sequence would be '2 *to the power of* 6', the 7th term would be '2 *to the power of* 7' and the 8th term '2 *to the power of* 8' and so on. Your children may also be able to predict terms out of order, such as the 10th term being '2 *to the power of* 10'; the 24th term being '2 *to the power of* 24' and the 100th term being '2 *to the power of* 100', and so on.

Again, by spotting patterns and being able to predict terms out of order, your children will have made the connection between the 'position number' and the relationship. In this example the 'position number' reappears as the power number in the relationship. So the 10th term is '2 *to the power of* 10', the 17th term is '2 *to the power of* 17', the 98th term being '2 *to the power of* 98' and so on.

The nth term is quite simply '2 *to the power of* n' and can be written as 2^n.

And this is the **formula** for the sequence.

Algebra

Using letter symbols to stand in for any number and creating formulae are the foundations of **algebra.** This is a significant and important branch of maths.

Algebra is a language and so it needs to be taught in a way that enables children to learn to speak it, write it and understand it – gradually. The very first steps of algebra will be introduced in primary school. Algebra should seem very easy and straightforward at this stage – because it is (honest!).

Children will be taught the basics of the language of algebra, starting with the following:

Letter symbols are used to represent **unknown** numbers.

Very simple algebra will probably have been used in class without anyone thinking about it. For example children may have been given number sentences to help them practise their four operations of number: adding, subtracting, multiplying and dividing. Sometimes it will be the answer that needs completing but sometimes it may be a different number that is missing. The missing number might be represented by a shape, a symbol or a letter.

Here are some examples, in which you need to 'find' the value of 'y':

$$4 + 5 = y \qquad \text{Answer: } y = 9$$
$$12 - y = 8 \qquad \text{Answer: } y = 4$$
$$y + 6 = 18 \qquad \text{Answer: } y = 12$$
$$3 \times 9 = y \qquad \text{Answer: } y = 27$$
$$27 \div 9 = y \qquad \text{Answer: } y = 3$$
$$42 \div y = 7 \qquad \text{Answer: } y = 6$$
$$y \div 7 = 8 \qquad \text{Answer: } y = 56$$

In all of the above, 'y' is a letter symbol being used to represent the 'missing' or unknown number.

If children have mastered the four operations of number and can add, subtract, multiply and divide with ease they will probably find all of the above very straightforward. If so, they are already using algebra!

Only if children are frightened of algebra will they freeze when presented with examples such as those above. If it is introduced naturally and confidently children will simply carry on as normal. If your children show any hesitation you can simply read the sum out loud:

- '4 plus 5 equals "what"?'
- '12 take away "what" equals 8?'
- … and so on.

Older children who *have* 'learnt' to be afraid of algebra *do* panic when faced with examples such as those above. They think it must be really hard and so often don't even try. This is a huge shame because often the questions are not difficult at all.

Most teachers know the following joke:

- In an exam one of the questions asks 'Find y'; the pupil responds by writing 'Here it is!' and drawing a big arrow next to the 'y' written in the question. Funny, yes. But too true, also.

Children panic and forget the simple stuff. We need to make sure they don't panic by being able to show them we are not afraid of algebra.

Here are some more basics of the language of algebra:

- n + n + n + n is the same as '4 lots of n', or 4 × n.
- 4 × n is conventionally shortened to 4n to avoid using the multiplication sign.

The multiplication sign (×) is traditionally *not* used in algebra because it can be so easily confused with the letter symbol (x). Instead we have what is called 'implied multiplication'. Four times 'n' *could* be written as '4 × n' but rewriting it without the multiplication sign avoids confusion and is a more elegant representation: '4n'. We just need to remember that '4n' *means* '4 × n'.

- $n \times 4$ is also rewritten as $4n$, because: $n \times 4$ is the same as $4 \times n$ and we always write the number first.
- $a \times b$ is similarly rewritten as ab; likewise '$a \times (b + c)$' is rewritten as '$a(b + c)$'.
- a^2 is the same as '$a \times a$' or 'a squared'.

A **very** common mistake is to think '$a \times a$' is the same as $2a$, which it is not. However, $a + a$ is the same as $2a$.

- '$a \times a \times a$' is the same as a^3 or 'a cubed'.

Again, a very common mistake is to think that '$a \times a \times a$' is the same as $3a$, which it is not. However, $a + a + a$ is the same as $3a$.

- $(3n)^2 = (3n) \times (3n) = 9n^2$

Note the use of brackets around the $3n$. Without the brackets $3n^2$ is just $3n^2$.

- When 'doing' algebra all the rules and conventions of normal arithmetic apply.

For example:

If $a = 2$ and $b = 5$ find: $3(a + b)$

Just as with normal arithmetic we do the 'bit' in the brackets first.

$(a + b)$ is equal to $(2 + 5)$ which equals 7
$3(a + b)$ equals 3×7
$3(a + b) = 21$

- When we replace a letter symbol with a number, as in the example above, it is called **substitution**. It is just like taking one footballer off the pitch and putting another one on in their place.

Here are some more examples.

> **If a = 3, b = 6 and c = 2, calculate:**
> $4 + a$ Answer: $4 + a = 7$
> $b + c$ Answer: $b + c = 8$
> $b - 5$ Answer: $b - 5 = 1$
> $a + b + c$ Answer: $a + b + c = 11$
> $b - c - 2$ Answer: $b - c - 2 = 2$
> $2a$ Answer: $2a = 6$
> ab Answer: $ab = 18$
> $12 - ac$ Answer: $12 - ac = 6$
> $a(b + c)$ Answer: $a(b + c) = 24$

- This is an example of an **expression**: $2a + 5$
- This is an example of an **equation**: $2a + 5 = 17$

The difference between an expression and an equation is that an equation has an equals sign, whereas an expression does not.

Children will be asked to write simple expressions…

(Often 'n' will be used to represent the unknown number but any lower-case letter can be used.)

For example:

- Add 5 to an unknown number   Answer: $n + 5$
- Subtract 3 from a number   Answer: $n - 3$
- Multiply a number by 10   Answer: $10n$
- Multiply a number by itself   Answer: n^2
- Divide a number by 4   Answer: $n \div 4$ or $n/4$
- Add 2 to a number and then
  multiply by 7   Answer: $7(n + 2)$
- Multiply a number by 2 and then
  add 5   Answer: $2n + 5$

Children will be asked to solve simple equations…

For example:

$a + 4 = 10$ Answer: $a = 6$

$5a = 15$	Answer: $a = 3$
$8a - 2 = 6$	Answer: $a = 1$
$2a + 5 = 17$	Answer: $a = 6$

At this stage, solving simple equations will be about 'having a go' – thinking about what the unknown number could be and then trying it out. Having a 'feel' for equations is a very significant starting point and it can help to demystify the whole process.

So, in the first example given above, '*"what" plus 4 equals 10?*' And in the second, '*5 lots of "what" equals 15?*' The third and the fourth are a little harder but again encourage your children to just have a go. Try any number for now and see if it 'works'. In the third example, start by replacing 'a' with 1...'*8 lots of 1 is 8 and take away 2 equals 6. Yes it works! So, "a" must equal 1*'.

Do the same for $2a + 5$. Start by trying with 4. For example: '*2 lots of 4 is 8, add 5 is 13. No, that's not right.*' Try again with 5... '*2 lots of 5 is 10, add 5 equals 15. No!*' And again, this time with 6... '*2 lots of 6 is 12, add 5 equals 17. Yes!! So a = 6.*'

Again, good arithmetic skills are essential to enable the focus to be on understanding the new ideas and not simultaneously stumbling over the basic maths.

- What does 'make a the subject' mean?

Basically it just means: rearrange the equation to find what 'a' equals. Bear in mind equations must always balance, in other words what you do to one side you must do to the other.

Here is an example:

$$2a - b = c \qquad \text{add } b \text{ to both sides}$$
$$\rightarrow \quad 2a = c + b \qquad \text{divide both sides by 2}$$
$$\rightarrow \quad a = \frac{c + b}{2}$$

(Note: the 'equals' signs *must always* be beneath one another.)

PROBLEM SOLVING USING ALGEBRA

Here is a 'wordy' type problem that can be solved using algebra:

- 'Lucy and her brother have 25 years of experience between them of working in interior design. Lucy's brother has 7 years more experience than her. How many years experience does Lucy have?'

Try it. You can obviously do this without algebra but I want to show you how quick – and easy – it is to solve using algebra.

If we say Lucy has 'n' years' experience, then her brother must have n + 7 years' experience (because he has 7 more years in interior design than Lucy).
    Now between them they must have

n + n + 7 years' experience.

From the question we know they have 25 years' experience between them.

So	n + n + 7	must equal 25
We can rewrite this as	n + n + 7	= 25
And again as	2n + 7	= 25
Now if we subtract 7 from both sides...	2n	= 18
Then halve both sides	n	= 9

Answer: 'Lucy has 9 years' experience (and her brother has 16 years' experience).'

'Think of a number' type questions are very popular in school. These are very simple to solve using algebra and involve children writing their own equations before solving them.

Here are some examples:

I think of a number and then subtract 8, the answer is 20. What is my number?

Answer: Let n be the number.

$$n - 8 = 20$$
$$n = 28$$

I think of a number and then add 12, the answer is 25. What is my number?

Answer: Let n be the number.

$$n + 12 = 25$$
$$n = 13$$

I think of a number, multiply it by 4 and then add 1, the answer is 9. What is my number?

Answer: Let n be the number.

$$4n + 1 = 9$$
$$n = 2$$

I think of a number, multiply it by 5 and then subtract 3, the answer is 17. What is my number?

Answer: Let n be the number.

$$5n - 3 = 17$$
$$n = 4$$

A typical question from typical teenagers, and some pre-teens, is:

- 'Why do we have to do algebra?'

Well, my stock answer (in age-appropriate language) goes something like this...

Algebra is such a prominent part of mathematics. It has a multitude of uses and is intrinsically perfect in its precise and efficient way of communicating ideas and solving problems.

Algebra *is* the language of mathematics. It is an accurate short-hand for recording ideas.

Algebra is widely used in problem solving and decision making even in everyday life. Simply altering a cooking recipe from a 'recipe for two' to a 'recipe for 5' involves algebraic calculations. Admittedly, you might not write down the formula as: $\frac{a}{2} \times 5$, where 'a' represents each quantity in the recipe, but you are nevertheless carrying out the calculation. In fact, many problems involving money, time, distance, area, volume and so on will use algebra.

It is 60 miles to the beach. I want to get there within 2 hours. What is the average speed I must travel?
Answer: 30 miles per hour, as Speed = Distance ÷ Time.

The cost of this carpet is £30 per square metre and my room is 8 metres long and 4 metres wide. How much will it cost?

First, we need to think about how we find the area of a rectangle, namely 'Area equals Length times Width' ($A = l \times w$), which is in itself an example of algebra. And then multiply the area (number of square metres) by the cost per square metre.
Answer: cost equals: $30 \times (8 \times 4)$.

But what if I don't know the dimensions of all the rooms I want to carpet? And what if I want to consider different carpets with different price tags? Now a general formula is more useful.
Answer: cost equals: $p \times (l \times w)$, where p is the price of the carpet per square metre, and l and w are the length and width of the room, respectively.

Many professions will rely on algebraic formulae to operate efficiently and effectively. They include carpenters, when calculating precise measurements; construction workers, when considering weight-bearing loads; medical practitioners, when calculating drug doses for patients, and many others.

Even sport uses algebra. There are several recognised types of intelligence, 'physical intelligence' being one. David Beckham is a classic example of pure physical intelligence at play. The judgement he makes when he kicks a ball relies on juggling many variables ('unknowns') such as the angle, the force of the kick and the estimated distance. I am sure he is unaware he is using algebra – but he is!

Handling variable quantities to solve problems, inform decisions or reach/restore a sense of balance (or score the perfect goal) is what it is all about.

Algebra is inherent in many other branches of mathematics and is of utmost importance to all areas of science. All science formulae are examples of algebra.

Think about all the formulae you can (vaguely) remember from school:

- $e = mc^2$, which is Einstein's famous equation relating mass to energy.
- $A = \pi r^2$, which expresses the relationship between the radius of a circle and the circumference.
- $C = \frac{5}{9}(F - 32)$, which is used to convert degrees Fahrenheit to degrees Celsius.

In all of these algebra is at play.

I know that little of the above would persuade the typical teenager of algebra's merits. So, I simply say... it is good for you, it strengthens the mind and encourages precise logical thinking – skills any employer would appreciate.

In the next chapter we learn some more useful rules, this time for working with fractions, decimals and percentages.

7.
FRACTIONS, DECIMALS AND PERCENTAGES

Fractions fill some of us with fear. They look complex, they have lots of little numbers and give us the anxious feeling that there are some rules about them that we were once taught but can't quite remember. If this is how you feel when your children ask for help with fractions, it might be an idea to go back to basics and remember what fractions actually *are*. Fractions are simply a 'bit' of a whole thing.

- 'Do you want your toast cut in half, or should I leave it whole?'

I bet we've all said something like this to a typically fussy 5 year old at breakfast time. And if the 5 year old fancies it in halves, then we cut the toast into 2 equal pieces. In this way we are already showing them that **2 halves** (of toast) **make a whole** (piece of toast).

UNDERSTANDING THE BASICS
(RECEPTION AND YEARS 1 AND 2, AGES 5–7)

Simple fractions of halves and quarters will be explored in the first few years of school. Finding **one half, one quarter** and **three quarters** of shapes and sets of objects is usually the goal by the end of Year 2.

Having said that, I think introducing other fractions in conversation is a really worthwhile exercise. Using fractions in everyday conversation, while children are still young, can mean they may naturally appreciate and understand the language of fractions without it becoming a big deal later on. Young children really do absorb new ideas like a sponge.

Thinking about our 'toast' scenario again, there is absolutely nothing to stop us from taking it further – as far as your children's interest allows. Once the toast is in quarters (4 pieces), you could ask your children:

- 'How many pieces will there be if we cut each quarter in half?'
- '8. Yes. These pieces are called eighths. So you need **8 eighths** to make **a whole** piece of toast.'

You might want to emphasise the 'ths' sound at the end of '*eighths*' as lots of fractions end in this sound.

- 'And what if we cut each of these eighths in half?'
- 'There are lots of little pieces now.' (Perfect soldiers for a boiled egg?)
- '16 of them and they are called sixteenths, so **16 sixteenths** make **a whole** piece of toast.'

This may be too far for your children, or not far enough. But remember, this is only conversation. We are not trying to make them 'learn' anything at the breakfast table, just to become subliminally familiar with the language of fractions.

Children will be taught at school how to recognise and write *simple* fractions numerically.

For example:

- One half   ½
- One quarter   ¼

The 'one' on the top is the **numerator** and can be seen to represent **'one whole'**. The **'whole'** can be a whole piece of toast, a whole cake, a whole pizza, a whole picture, a whole set of buttons, all the children in a class. It doesn't matter, it just represents the whole thing, the full amount, all of it, the total. (You may be performing a mental leap here and asking yourself, what happens when the numerator is a higher number than one? Bear with me here; your question is answered later, but let's stick with this definition for now.)

The number on the bottom is called the **denominator** and shows how many pieces the 'whole' has been cut or divided into.

So ½ shows that the '1' **whole** has been cut into '2' equal pieces: 1 divided by 2.

The ¼ shows that the '1' **whole** has been cut into '4' equal pieces: 1 divided by 4.

Children may be shown pictures like the ones below and asked what fraction has been shaded.

Other questions that children are likely to hear at school are:

- 'Can you tell me: what is half of 8?'
- 'What is half of 12?'
- 'What is half of 18?'

At this stage, children will commonly be asked to halve *any even number* up to 20. Beyond this, they may also be asked to become familiar with the following:

- 'Can you tell me: what is half of 50?'
- 'What is half of 100?'

In each of the examples above, to find the answer it is a matter of looking at the **whole** amount and cutting or sharing it in two. The symbol for sharing or dividing, is '÷' as seen in Chapter 5: Division. But '/' and '–' are also symbols to show division.

So the answers to the above questions are:

- ⁸⁄₂ is 4   which you would read as   *'8 divided by 2 is 4'*
- ¹²⁄₂ is 6   which you would read as   *'12 divided by 2 is 6'*
- ¹⁸⁄₂ is 9   which you would read as   *'18 divided by 2 is 9'*
- ⁵⁰⁄₂ is 25   which you would read as   *'50 divided by 2 is 25'*
- ¹⁰⁰⁄₂ is 50   which you would read as   *'100 divided by 2 is 50'*

Questions related to quarters would be along the lines of:

- 'Can you find a quarter of 8?'
- 'I have 12 buttons. How many would a quarter of them be?'
- 'I am going to share this cake between the 4 teddy bears. What fraction would each teddy bear get?'
- 'What is a quarter of 100?'

In each of these examples, it would be a matter of looking at the **whole** amount and cutting or sharing it in 4.

So the answers are:

- $\frac{8}{4}$ is 2   which you would read as   *'8 divided by 4 is 2'*
- $\frac{12}{4}$ is 3   which you would read as   *'12 divided by 4 is 3'*
- 1 cake shared by 4: which you would read as *'¼ (one quarter)'*
- $\frac{100}{4}$ is 25 which you would read as   *'100 divided by 4 is 25'*

So children will learn that **one whole** can be divided into (or shared between, broken into, split into, cut into… any of these words might be used) '*2 identical halves*' or '*4 identical quarters*'. And that: '*2 halves make one whole*', and '*4 quarters make one whole*'.

Another easy way to show this (and slightly less messy than using buttered toast) is to take a piece of paper and fold it exactly in half. Open it up and see there are 2 parts the same (2 halves). Fold it up again and then fold exactly in half again. Open it up to see there are now 4 identical parts (4 quarters).

Children may also begin to recognise that 2 quarters are the same as one half. They *may* also appreciate that one half and one quarter is the same as three quarters.

MOVING FORWARD
(YEARS 3 AND 4, AGES 7–9)

In the early years children will have learnt these hugely important rules:

- Finding a half of 'something' is the same as dividing by 2.
- Finding a quarter of 'something' is the same as dividing by 4.

Now, as well as recognising halves and quarters, children will be introduced to other fractions, namely thirds, fifths and tenths.

- One third $\frac{1}{3}$
- One fifth $\frac{1}{5}$
- One tenth $\frac{1}{10}$

These are all known as **unit fractions**. This means that a unit (the number 1) is the **numerator** (the top number) and the **denominator** (the bottom number) can be any number you like.
So:

- $\frac{1}{7}$ is one seventh, and means **I whole** divided or cut into 7 equal pieces.
- $\frac{1}{8}$ is one eighth, and means **I whole** divided or cut into 8 equal pieces.
- $\frac{1}{12}$ is one twelfth, and means **I whole** divided or cut into 12 equal pieces.
- $\frac{1}{20}$ is one twentieth, and means **I whole** divided or cut into 20 equal pieces.

Knowing and understanding the above can help children avoid a very common mistake.

Question: Which is bigger $\frac{1}{12}$ or $\frac{1}{8}$?
Answer: $\frac{1}{8}$ is bigger than $\frac{1}{12}$.
$\frac{1}{12}$ is one bit of a whole piece of toast (or whatever) after it has been cut into 12 pieces.
$\frac{1}{8}$ is one bit of a whole piece of toast (or similar) after it has been cut into 8 pieces. Each of these will be a bigger bit of toast than $\frac{1}{12}$.

Why do so many children get in a muddle with this? In my experience, these are the children who have not fully understood the basics of fractions, often because they have moved on too quickly, and so are less confident in interpreting fractions. Instead children in this situation understandably rely on maths they are familiar with. They see a number 12 and a number 8 and make the *wrong* assumption. They think the 'thing' with a 12 in it must be bigger than the 'thing' with an 8 in it.

The language of fractions is interesting and some children will notice apparent discrepancies. If we cut or share or divide something into 2 identical pieces we have **halves**. Cut something into 3 identical bits and we have **thirds**. Cut something into 4 identical parts and we have **quarters.**

Into 5 pieces? Then we have **fifths**. Into 6 pieces? Then it is **sixths**. Likewise for **sevenths, eighths, ninths, tenths** and so on.

So why 'halves', 'thirds' and 'quarters'? Why not 'twoths', 'threeths' and 'fourths'? Lots of children ask this question, and of course these language traditions are so old that no one really knows. Sometimes just knowing '*it is because it is*' helps organise their thoughts and reassures them.

Later, children will be introduced to other types of proper fractions.

Proper fractions are fractions where the **numerator** (the number on the top) is smaller than the **denominator** (the number on the bottom). For example, ⅚ is a proper fraction. But fractions with a bigger numerator than denominator are called **improper fractions** (or 'top-heavy' fractions), such as ⅚, of which more later.

All of these fractions are also called **vulgar** or **common** fractions.

Going back to our earlier description of the numerator as being 'the whole', sometimes a leap of learning needs to take place at this stage.

'Two thirds' looks like ⅔. Following the logic set out earlier in this chapter, ⅔ would represent **two whole** pieces (of toast, say) shared into 3 equal portions. And indeed this is correct.

But what it also means, and how it is usually interpreted, is exactly what it sounds like if you say the words out loud, '*two thirds*'. That is 2 lots of '*one third*'. So if you had cut your toast into thirds (3 equal pieces), two thirds (⅔) would be 2 of these pieces.

However, if you preferred the first interpretation of two thirds (⅔) as being 2 whole pieces of toast shared into 3 equal portions, you would still end up with exactly the same amount of toast on your plate!

'Equivalent fractions' is the name given to fractions that are the same.

One half of a piece of toast is exactly the same as two quarters of a piece of toast. So 'one half' and 'two quarters' are equivalent fractions:

$$\tfrac{1}{2} = \tfrac{2}{4}$$

Other commonly used equivalent fractions are:

- 'four eighths' and 'one half' $\tfrac{4}{8} = \tfrac{1}{2}$
- 'four eighths' and 'two quarters' $\tfrac{4}{8} = \tfrac{2}{4}$
- 'five tenths' and 'one half' $\tfrac{5}{10} = \tfrac{1}{2}$
- 'two halves' and 'one whole' $\tfrac{2}{2} = 1$
- 'four quarters' and 'one whole' $\tfrac{4}{4} = 1$
- 'five fifths' and 'one whole' $\tfrac{5}{5} = 1$

It is really useful for children to recognise that whenever the numerator and denominator are the same number, the fraction itself is equivalent to **'1 whole'** (as in the last three examples above).

Below are some more commonly used equivalent fractions:

$$\tfrac{2}{8} = \tfrac{1}{4}$$
$$\tfrac{6}{8} = \tfrac{3}{4}$$
$$\tfrac{2}{10} = \tfrac{1}{5}$$
$$\tfrac{2}{6} = \tfrac{1}{3}$$
$$\tfrac{4}{6} = \tfrac{2}{3}$$

The Number Line will undoubtedly be used to demonstrate the positioning and magnitude (size) of fractions to your child. Starting at 0 and ending at 1, various proper fractions can be shown.

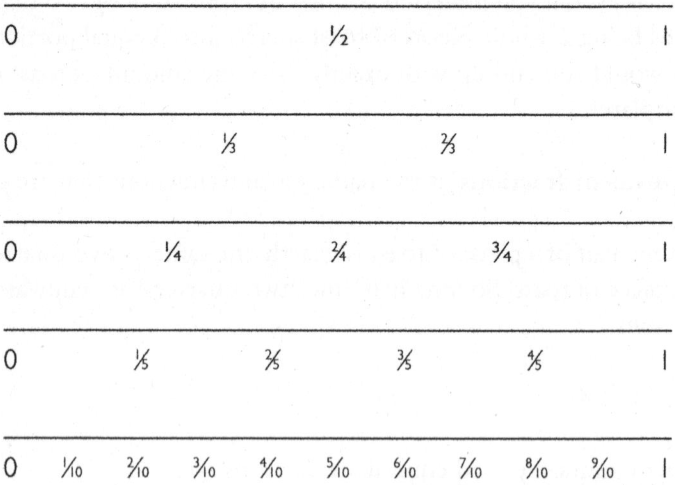

| 0 | | ½ | | 1 |

| 0 | ⅓ | | ⅔ | | 1 |

| 0 | ¼ | | 2/4 | | 3/4 | | 1 |

| 0 | ⅕ | | ⅖ | | ⅗ | | ⅘ | | 1 |

| 0 | ⅒ | 2/10 | 3/10 | 4/10 | 5/10 | 6/10 | 7/10 | 8/10 | 9/10 | 1 |

Diagrams, similar to the one above (where a series of Number Lines are shown simultaneously), are called '**fraction walls**'. Below is a classic example of a fraction wall. Each 'layer' of the wall is divided into pieces to illustrate different fractions. In this example, working top down, the layers show: one whole, halves, thirds, quarters, sixths, ninths, tenths and twelfths respectively.

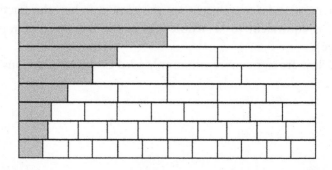

Here is another example to show a different set of fractions:

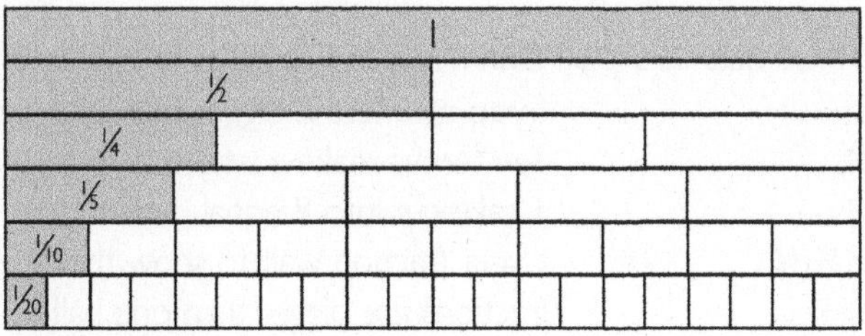
1
½
¼
⅕
1/10
1/20

Children will use these in class as props to help them see how fractions relate to each other.

For example:

- Equivalent fractions can be identified ($\frac{1}{2} = \frac{2}{4} = \frac{5}{10}$).
- The size of relative fractions is also made clear – 'seven tenths' ($\frac{7}{10}$) is bigger than 'two thirds' ($\frac{2}{3}$).

Questions children are likely to hear at school are:

- 'What is $\frac{1}{10}$ of 30?'
- 'What is three quarters of 20?'
- 'What number is halfway between 4 and 5?'
- 'Which is bigger: one half or three quarters?'
- 'This cake has been cut into 10 equal slices. We can see 7 pieces so $\frac{7}{10}$ of the cake is left. What fraction has been eaten?'
- 'At Thea's birthday party the whole cake is divided equally between 8 children. What fraction of the cake do they each get?'
- 'Which of these fractions are bigger than one half:
  $\frac{2}{3}$ • $\frac{6}{10}$ • $\frac{1}{4}$ • $\frac{2}{5}$ • $\frac{5}{8}$'

The answers to the above questions are:

3	30 is shared or divided into 10 equal parts: $\frac{30}{10} = 3$
15	First 20 is shared or divided into 4 equal parts (quarters). $\frac{20}{4} = 5$. Each quarter is 5. So three quarters is '3 lots of 5' which is 15.
$4\frac{1}{2}$	

Three quarters (¾)	One half is the same as two quarters, and three quarters is bigger than two quarters
³/₁₀	Ten tenths make a whole: $\frac{7}{10} + \frac{3}{10} = \frac{10}{10}$
⅛	1 cake cut into 8 equal pieces
⅔ ⁵/₁₀ ⅝	Use a 'fraction wall' to show that these fractions are bigger than one half

Other typical questions begin '*What fraction of...*'
For example:

- 'What fraction of £1 is 50p?'
- 'What fraction of £1 is 20p?'
- 'What fraction of 1 metre is 10cm?'
- 'What fraction of 1 metre is 75cm?'
- 'What fraction of the larger bag of potatoes is the smaller bag of potatoes?'   2 kg, 7 kg
- 'What fraction of the bigger shape is the smaller one?'

     * * *   * * * * * * *

In all these examples above, we are trying to find the smaller amount as a fraction of the larger amount.

⬤ What fraction of £1 is 50p?
At this stage it will be assumed that children know two 50p coins are needed to make £1. Therefore one 50p is a half of £1.
Answer: ½

But there is an alternative approach:

⬤ There are 100p in a pound.
50p out of 100p can be written as the fraction ⁵⁰/₁₀₀.
And ⁵⁰/₁₀₀ is equivalent to ⁵/₁₀. (See the next section for more on 'equivalent fractions'.) And ⁵/₁₀ is equivalent to ½.
Answer: ½

🌑 What fraction of £1 is 20p?
   Again it will be assumed that children know five 20p coins are
   needed to make £1. One 20p is therefore 1 out of 5 coins or
   one fifth of £1.
   Answer: ⅕

But just as above, there is an alternative approach:

🌑 There are 100p in a pound.
   20p out of 100p can be written as the fraction ²⁰⁄₁₀₀.
   ²⁰⁄₁₀₀ is equivalent to ²⁄₁₀.
   And ²⁄₁₀ is equivalent to ⅕.
   Answer: ⅕

🌑 What fraction of 1 metre is 10cm?
   Children will have worked a lot with centimetres and metres.
   They will be familiar with the fact that 100cm make one metre.
   They will probably have practised splitting a metre into 10 lots of
   10cm. One lot of 10cm is therefore one tenth of a metre.
   Answer: ¹⁄₁₀

The alternative approach:

🌑 10cm out of 100cm can be written as the fraction ¹⁰⁄₁₀₀.
   ¹⁰⁄₁₀₀ is equivalent to ¹⁄₁₀.
   Answer: ¹⁄₁₀

🌑 What fraction of 1 metre is 75cm?
   50cm is the same as ½ a metre.
   25cm is the same as ¼ of a metre.
   So 75cm is the same as ½ a metre plus a ¼ of a metre. And ½
   plus ¼ = ¾.
   Answer: ¾

Alternatively:

⬢ 75cm out of 100cm can be written as the fraction $^{75}/_{100}$.
$^{75}/_{100}$ is equivalent to ¾.
Answer: ¾

⬢ 'What fraction of the larger bag of potatoes is the smaller bag of potatoes?' 2 kg, 7 kg
Answer: '2 out of 7' and is written as the fraction $^2/_7$.

⬢ 'What fraction of the bigger shape is the smaller one?'
    * * *   * * * * * * * *
Answer: '3 out of 8' and is written as the fraction $^3/_8$.

Mixed numbers is the name given to a number made up of a whole number part and a fractional part.
Examples are:

5½	'five and a half'
2¼	'two and a quarter'
10¾	'ten and three quarters'

Mixed numbers can be converted into **improper fractions**. Let's look at each of the above in turn:

      5½

Think of 5½ apples. We want to give 'half' an apple to each child at play time. How many children could we give 'half' an apple to? In other words, how many 'halves' do we have?

- 1 apple would give us 2 'halves';
- 2 apples would give us 4 'halves';
- 3 apples would give us 6 'halves';
- 4 apples would give us 8 'halves';

- 5 apples would give us 10 'halves';
- 5½ apples would give us 11 'halves'.
- So 5½ is the same as ¹¹⁄₂ (11 halves).

¹¹⁄₂ is an **improper fraction** because the numerator (top number) is bigger than the denominator (bottom number).

2¼
- 1 is the same as 4 quarters;
- 2 is the same as 8 quarters;
- 2¼ is the same as 9 quarters.
- So 2¼ is the same as ⁹⁄₄ (9 quarters).

10¾
- 1 is the same as 4 quarters;
- 10 is the same as 40 quarters;
- 10¾ is therefore the same as 43 quarters.
- So 10¾ is the same as ⁴³⁄₄ (43 quarters).

Mixed numbers can also be shown on a Number Line. For example 10¾ is shown below. This helps reinforce the idea that 10¾ lies *between* 10 and 11 and is closer to 11 than 10.

10	10¾	11

Decimal notation

Children will already have come across **Decimal notation** when looking at place value with 'tenths' and 'hundredths' (see also page 34).

Tenths are, of course, fractions just the same as eighths or ninths are – they are a whole divided into 10 equal parts. One tenth = ¹⁄₁₀.

We are familiar with working in hundreds, tens, units and so on, and tenths fit neatly into this decimal number system, separated from whole numbers by the decimal point. When a tenth is represented using this system it is referred to as a **decimal fraction**. (The same is true of hundredths and thousandths, of course, but the link between tenths and other familiar fractions may be easier to grasp at this stage.)

As a quick recap…

H	T	U	.	t	h
Hundreds	Tens	Units	Decimal point	tenths	hundredths
100	10	1	.	$\frac{1}{10}$	$\frac{1}{100}$
		0	.	1	

The number '0.1' means **no units** and **one tenth**; '0.1' is the decimal fraction of 'one tenth'.

> The term 'decimal fractions' is often just shortened to 'decimals'.

If your children are happy with '0.1' being exactly the same as '$\frac{1}{10}$', they may also appreciate other decimal fractions and that:

0.2	is equivalent to	$\frac{2}{10}$
0.3	is equivalent to	$\frac{3}{10}$
0.4	is equivalent to	$\frac{4}{10}$
0.5	is equivalent to	$\frac{5}{10}$
0.6	is equivalent to	$\frac{6}{10}$
0.7	is equivalent to	$\frac{7}{10}$
0.8	is equivalent to	$\frac{8}{10}$
0.9	is equivalent to	$\frac{9}{10}$
1.0	is equivalent to	$\frac{10}{10}$

Your children will also find it essential to know, and may just need to *learn*:

0.5	is equivalent to	$\frac{1}{2}$

0.25	is equivalent to	¼
0.75	is equivalent to	¾
0.1	is equivalent to	⅒

TO THE TOP
(YEARS 5 AND 6 PLUS, AGES 9–12 PLUS)

As children progress through the school years, they will continue to consolidate all of the knowledge covered in the earlier sections of this chapter. Coming back to fractions after a break, you may notice that your children have apparently 'forgotten' everything they used to know about them. This is not unusual. Fractions are not something most children encounter on a daily basis, so regular revision of the rules we've learned is essential. You may wish to revisit earlier sections of this chapter before helping older children with this subject.

GETTING FRACTIOUS?

Fractions can cause angst, there's no doubt about that from my own experience. The expectation that fractions are going to be tricky may not help. But like all other areas of maths, there are logical rules to follow. With these rules firmly in place fractions should pose few problems.

Perhaps the 'point' of fractions related to everyday life is not obvious enough. Certain careers will require a working knowledge of fractions, but what about the rest of us? In fact, we all deal with them on a daily basis, but perhaps not always consciously. For example:

- 'How much do I save on the 3-for-2 offer?'
- 'This is half price in the sale.'
- 'These jeans are a third off.'
- 'I've finished three walls of the room, will I have enough paint left?'

- 'If I keep the car another year, how much of its value will it lose?'
- 'If everyone at dinner will drink 3 glasses of wine, how much should I buy?'

So there's more to fractions than toast, but in reality many of us may be more familiar with decimals and percentages. Now that inches and ounces are on the way out, our weights, measures, currency and interest rates all lend themselves most readily to the decimal system of tenths, hundredths and thousandths.

We just need to remember that decimal numbers and percentages are alternative forms of common (or vulgar) fractions. They are all the same thing.

But an advantage of fractions (over decimals and percentages) is the way they can express a value exactly. Decimals and percentages can often require crude approximating. Compare $\frac{3}{7}$ with its decimal equivalent of 0.428571...

By now children will have been introduced to: **proper fractions, improper fractions, mixed numbers**, the terms '**numerator**' and '**denominator**', **fraction walls, equivalent fractions, finding fractions, expressing one number as a fraction** of another, and begin to appreciate that fractions and decimal numbers are **related**.

This is not to say that children will have understood or 'learnt' it all. They are likely to need much more practice over many more years before they have grasped the subject completely.

As we have seen in the previous sections, fractions can be *read* or *interpreted* in several different ways.

For example: $\frac{5}{8}$ can be read or pictured as:

- 'five eighths',
- '5 divided by 8',
- '5 out of 8 (as in a test)',
- '5 pieces of a pie, when the pie has been cut into 8 equal pieces',

- '5 pies shared equally between 8 people',
- '… as the same as $\frac{10}{16}$ or $\frac{15}{24}$ or $\frac{20}{32}$ or $\frac{25}{40}$…' (or any equivalent fraction).

Equivalent fractions were introduced in 'Moving Forward' (see page 309). Equivalent fractions have exactly the same value as each other. So if you ate $\frac{5}{8}$ of a pie, you would be eating exactly the same amount as if you'd had $\frac{10}{16}$ of a pie. (Just 5 bigger slices as opposed to 10 skinny ones.)

Initially, children will be taught how to identify equivalent fractions using a **fraction wall** or similar (see page 310).

Later, children will be shown that multiplying (or dividing) both the numerator *and* the denominator by the *same* amount will produce equivalent fractions.

Looking at $\frac{5}{8}$ again:

- Multiply both numerator and denominator by 2 and we get $\frac{10}{16}$.
- Multiply both numerator and denominator by 3 and we get $\frac{15}{24}$.
- Multiply both numerator and denominator by 4 and we get $\frac{20}{32}$.
  …And so it goes.

This method works for all fractions, both **proper** and **improper**.

- $\frac{5}{8}$ is a **proper** fraction because the numerator is smaller than the denominator.
- $\frac{8}{5}$ (*eight fifths*) is an **improper** fraction because now the numerator is bigger than the denominator.

Improper fractions can be converted to **mixed numbers**.

Mixed numbers are so called because they are made up from a mixture of whole numbers and fractional parts.

- $\frac{8}{5}$ is the same as $1\frac{3}{5}$ (*one and three fifths*).

Why? Because it takes $\frac{3}{3}$ to make **1 whole** and then there are $\frac{1}{3}$ left over.

- $\frac{17}{3}$ is the same as $5\frac{2}{3}$ (*five and two thirds*).

Why? Because it takes $\frac{3}{3}$ to make **1 whole**, $\frac{6}{3}$ to make **2 wholes**, $\frac{9}{3}$ to make **3 wholes**, $\frac{12}{3}$ to make **4 wholes** and $\frac{15}{3}$ to make **5 wholes**. Then there are $\frac{2}{3}$ left over.

In the previous section we looked at expressing one number as a fraction of another.

For example:

What fraction of the larger bag of potatoes is the smaller bag of potatoes? 2 kg, 7 kg
The **answer** is '2 out of 7' and is written as the fraction $\frac{2}{7}$.

Now children will also be expected to '*express a larger whole number as a fraction of a smaller one*'.

For example:

Peter's favourite pizzas come from 'Perfect Pizzas' and each one is pre-cut into 5 slices. What fraction of a pizza is 7 slices?
Answer: Each slice of pizza is $\frac{1}{5}$ so 7 slices of pizza equals $\frac{7}{5}$.
$\frac{7}{5}$ is an improper fraction and can also be written as a mixed number of $1\frac{2}{5}$ (or 1 whole pizza and 2 slices of another pizza).

Simplifying fractions by cancelling common factors is one of those grandly named methods that can sound a bit scary, but really isn't. It's something your children will be taught to help them express their answers in the simplest and most easily understood way. It also helps children revisit equivalent fractions, and introduces common factors and denominators, useful later in adding and subtracting fractions (more below).

Looking at an example will help.

⁴⁄₂₀ (*four twentieths*) can be **reduced or cancelled down**. This simply means finding an equivalent fraction for ⁴⁄₂₀ with smaller numbers for the numerator and for the denominator.

We start by halving the numerator and denominator and then halving again:

⁴⁄₂₀ = ²⁄₁₀ = ⅕.
So ⅕ is the simplified form of ⁴⁄₂₀.

This method is fine for this example. But what if you can't halve. For example, if either the numerator or denominator are not even? A more reliable way would have been to look at ⁴⁄₂₀ and think of a number that would go *into* both 4 and 20 (in other words find a **common factor** for both 4 and 20). In this case '4' is a common factor.

Dividing both the numerator and the denominator by 4 gives:

⁴⁄₂₀   =   ⅕

This is what is meant by 'simplifying fractions by cancelling common factors' (see also page 251).

Here is another example:

⬢ Simplify ³⁄₁₂.
Start by trying to find a common factor for 3 and 12. In other words, what 'goes into' both 3 and 12.
3 'goes into' both 3 and 12.
So we divide both numerator and denominator (top and bottom numbers) by 3.
(3 ÷ 3 = 1; 12 ÷ 3 = 4)
Answer: ³⁄₁₂ = ¼

A further example:

⬢ Reduce $^{15}\!/_{25}$ to its simplest form.
Again, we start by looking for a common factor for 15 and 25. In the previous examples the numerator was the common factor. This is not so in this example, so we have to look a little harder. What number 'goes into' both 15 and 25? Well, 5 does.
So we divide both numerator and denominator (top and bottom numbers) by 5.
$(15 \div 5 = 3; 25 \div 5 = 5)$
Answer: $^{15}\!/_{25} = \frac{3}{5}$

Converting fractions to have a common denominator is something children will need to be able to do, in order to put fractions in order of size and later to add and subtract fractions.
A typical question is:

- Place the following in order, smallest first:
$\frac{3}{5}$ • $^7\!/_{10}$ • $\frac{1}{2}$ • $\frac{3}{4}$

Converting fractions to have a common denominator is all a matter of using our knowledge of equivalent fractions. Encourage your children to write out a list of equivalent fractions for each of the fractions listed. To do this logically and efficiently (and to make sure they don't miss any out) get them to start by multiplying each numerator and denominator by 2.

$$\frac{3}{5} \quad = \quad ^6\!/_{10} \quad = \ldots$$
$$^7\!/_{10} \quad = \quad ^{14}\!/_{20} \quad = \ldots$$
$$\frac{1}{2} \quad = \quad \frac{2}{4} \quad = \ldots$$
$$\frac{3}{4} \quad = \quad ^6\!/_8 \quad = \ldots$$

Then continue by multiplying each numerator and denominator (of the original fraction) by 3, then by 4, then by 5, then by 6, and so on.

$$\frac{3}{5} = ^6\!/_{10} = ^9\!/_{15} = ^{12}\!/_{20} = ^{15}\!/_{25} = \ldots$$
$$^7\!/_{10} = ^{14}\!/_{20} = ^{21}\!/_{30} = \ldots$$

$$\tfrac{1}{2} = \tfrac{2}{4} = \tfrac{3}{6} = \tfrac{4}{8} = \tfrac{5}{10} = \tfrac{6}{12} = \tfrac{7}{14} = \tfrac{8}{16} = \tfrac{9}{18} = \tfrac{10}{20} = \ldots$$

$$\tfrac{3}{4} = \tfrac{6}{8} = \tfrac{9}{12} = \tfrac{12}{16} = \tfrac{15}{20} = \tfrac{18}{24} = \ldots$$

Now it's simply a matter of looking at each of the lists and considering whether there is a denominator common to them all. If not, you may need to continue the list a little longer. In the case above, they all have equivalent fractions with 'twentieths':

$$\tfrac{3}{5} = \tfrac{12}{20}$$
$$\tfrac{7}{10} = \tfrac{14}{20}$$
$$\tfrac{1}{2} = \tfrac{10}{20}$$
$$\tfrac{3}{4} = \tfrac{15}{20}$$

Now it is easy to put them in order starting with the smallest:

$$\tfrac{10}{20} \quad \bullet \quad \tfrac{12}{20} \quad \bullet \quad \tfrac{14}{20} \quad \bullet \quad \tfrac{15}{20}$$

Answer: $\tfrac{1}{2}$ • $\tfrac{3}{5}$ • $\tfrac{7}{10}$ • $\tfrac{3}{4}$

Adding and subtracting fractions: this is not particularly difficult, but several stages are often required and this can lead to mistakes. In the early years children may add very simple fractions in their head.

Examples of these might be:

$\tfrac{1}{4} + \tfrac{1}{4}$ which equals $\tfrac{2}{4}$ or $\tfrac{1}{2}$

$\tfrac{1}{2} + \tfrac{1}{4}$ which equals $\tfrac{3}{4}$

$\tfrac{1}{3} + \tfrac{1}{3}$ which equals $\tfrac{2}{3}$

Later, with more difficult sums, they may need pencil and paper. For example:

$\tfrac{2}{5} + \tfrac{1}{3}$

Often children just do not know where to begin when faced with a sum like this. But the starting point is always the same:

- To add or subtract fractions, the denominators *must* be the same.

- If the denominators are not the same then **equivalent fractions** with **common denominators** must be found first.

The denominators in our example ($\frac{2}{5} + \frac{1}{3}$) are different, so we can't add them yet. Instead we must first look for equivalent fractions. We do this as described above, multiplying each numerator and denominator by 2, and then by 3, 4, 5, 6 and so on.

- Equivalent fractions for $\frac{2}{5}$ are: $\frac{4}{10}$ $\frac{6}{15}$ $\frac{8}{20}$ $\frac{10}{25}$ $\frac{12}{30}$ ...
- Equivalent fractions for $\frac{1}{3}$ are: $\frac{2}{6}$ $\frac{3}{9}$ $\frac{4}{12}$ $\frac{5}{15}$ $\frac{6}{18}$ ...

Now we need to look to see whether we have gone far enough to find a common denominator (if not, keep writing down more equivalent fractions).

They both have an equivalent fraction with 'fifteenths':

$$\frac{2}{5} = \frac{6}{15}$$
$$\frac{1}{3} = \frac{5}{15}$$

Now we can replace the sum: $\frac{2}{5} + \frac{1}{3}$

with: $\frac{6}{15} + \frac{5}{15}$

This is now very easy, especially if you say the sum out loud.

'Six fifteenths add five fifteenths.'
Answer: 'Eleven fifteenths.'
$$\frac{6}{15} + \frac{5}{15} = \frac{11}{15}$$
So $\frac{2}{5} + \frac{1}{3} = \frac{11}{15}$

A common mistake that a lot of children make when adding (or subtracting) fractions is to think they have to add (or subtract) the denominators. This is just plain wrong! As long as the denominators are the same, all that is required is to simply add (or subtract) the numerators.

Another example:

$\frac{3}{8} + \frac{1}{3}$

The denominators in this example ($\frac{3}{8} + \frac{1}{3}$) are again different, so we can't add them yet. We must first look for and list equivalent fractions.

- Equivalent fractions for $\frac{3}{8}$ are: $\frac{6}{16}$ $\frac{9}{24}$ $\frac{12}{32}$ $\frac{15}{40}$ $\frac{18}{48}$ ...
- Equivalent fractions for $\frac{1}{3}$ are: $\frac{2}{6}$ $\frac{3}{9}$ $\frac{4}{12}$ $\frac{5}{15}$ $\frac{6}{18}$ $\frac{7}{21}$ $\frac{8}{24}$ ...

Now we need to look to see whether we have gone far enough to find a common denominator. They both have an equivalent fraction with 'twenty fourths':

$\frac{3}{8} = \frac{9}{24}$
$\frac{1}{3} = \frac{8}{24}$

Now we can replace the sum:	$\frac{3}{8} + \frac{1}{3}$	
with:	$\frac{9}{24} + \frac{8}{24}$	
	$\frac{9}{24} + \frac{8}{24} = \frac{17}{24}$	
Answer:	$\frac{3}{8} + \frac{1}{3} = \frac{17}{24}$	

And another example, this time with subtraction:

$\frac{5}{6} - \frac{2}{9}$

Again, the denominators are different so we can't subtract them yet.

- Equivalent fractions for $\frac{5}{6}$ are: $\frac{10}{12}$ $\frac{15}{18}$ $\frac{20}{24}$ ...
- Equivalent fractions for $\frac{2}{9}$ are: $\frac{4}{18}$ $\frac{6}{27}$ $\frac{8}{36}$ ...

They both have an equivalent fraction with 'eighteenths':

$\frac{5}{6} = \frac{15}{18}$
$\frac{2}{9} = \frac{4}{18}$

Now we can replace the sum: $\frac{5}{6} - \frac{2}{9}$

with: $\frac{15}{18} - \frac{4}{18}$

$\frac{15}{18} - \frac{4}{18} = \frac{11}{18}$

Answer: $\frac{5}{6} - \frac{2}{9} = \frac{11}{18}$

MULTIPLYING AND DIVIDING FRACTIONS

Multiplying

This is probably the easiest thing to do with fractions.

We simply multiply the two numerators together and then multiply the two denominators together.

For example:

$$\frac{2}{5} \times \frac{4}{9} = \frac{2 \times 4}{5 \times 9} = \frac{8}{45}$$

It's that easy!

With mixed numbers, simply convert them to improper fractions first and then continue as above.

For example:

$$2\tfrac{1}{4} \times 3\tfrac{1}{2}$$

is the same as:

$$\frac{9}{4} \times \frac{7}{2}$$

$$\frac{9}{4} \times \frac{7}{2} = \frac{9 \times 7}{4 \times 2} = \frac{63}{8}$$

You may then wish to convert the answer back into a mixed number:

$$\frac{63}{8} = 7\frac{7}{8}$$

Dividing

For example:

$$\frac{2}{5} \div \frac{4}{9}$$

Children will be taught to do this by following these rules:

- Leave the first fraction alone.
- Replace the division symbol (÷) with a multiplication symbol (×).
- Turn the second fraction upside down.
- Now multiply the fractions to find the answer.

So the above sum becomes:

$$\frac{2}{5} \div \frac{4}{9} = \frac{2}{5} \times \frac{9}{4} = \frac{2 \times 9}{5 \times 4} = \frac{18}{20}$$

This can be simplified to $\frac{9}{10}$ by halving both numerator and denominator.

Answer: $\frac{9}{10}$

Some children make the mistake of thinking there must be more to it and that you have to turn the answer upside down again. You don't.

Typical 'wordy' type questions can often catch children out. For example:

- 'What is ⅒ of ⅒?'
- 'Write ¾ of ⅖.'
- 'What is ⅛ of ½?'

These are so simple to answer once children remember the interchangeable use of the multiplication symbol 'x' and the word 'of'.

Now the questions become:

- 'What is ⅒ × ⅒?'
- 'Write ¾ × ⅖.'
- 'What is ⅛ × ½?'

Using the method of multiplication outlined in the above box:

$$\frac{1}{10} \times \frac{1}{10} = \frac{1}{100}$$

$$\frac{3}{4} \times \frac{2}{5} = \frac{6}{20} = \frac{3}{10}$$

$$\frac{1}{8} \times \frac{1}{2} = \frac{1}{16}$$

A very common misconception is that: It doesn't!	$0.1 \times 0.1 = 0.1$
This misconception may be so common because:	$1 \times 1 = 1$
To arrive at the correct answer...	
We know 0.1 is one tenth and can be written as ⅒.	
So 0.1 × 0.1 can be rewritten as:	⅒ × ⅒
And multiplying as above:	⅒ × ⅒ = ¹⁄₁₀₀
And ¹⁄₁₀₀ is the same as 0.01, so	$0.1 \times 0.1 = 0.01$

FRACTIONS, DECIMALS AND PERCENTAGES

Multiplying a whole number by a fraction

For example:

$\frac{1}{3} \times 12$

This may seem much easier if we remember the interchangeable use of 'x' and the word 'of'.

So, $\frac{1}{3} \times 12$ is the same as $\frac{1}{3}$ of 12. And $\frac{1}{3}$ of $12 = 4$

And another:

$16 \times \frac{1}{4}$

We know it doesn't matter in what order we multiply.

So, $16 \times \frac{1}{4}$ is the same as $\frac{1}{4} \times 16$. And $\frac{1}{4}$ of $16 = 4$

Alternatively, to multiply a **whole** number by a fraction, we simply turn the **whole** number into a top-heavy fraction and then multiply the fractions together, as shown in the box on page 326. To convert any **whole** number into a fraction simply divide it by one. (Dividing a number by 1 always leaves a number unchanged.)

The number 5 can be rewritten as:

$\frac{5}{1}$ or as $\frac{5}{1}$

So for the example:	$5 \times \frac{2}{3}$
We would rewrite this as:	$\frac{5}{1} \times \frac{2}{3}$
or again as:	$\frac{5}{1} \times \frac{2}{3}$
This is then easily solved:	$\dfrac{5 \times 2}{1 \times 3} = \dfrac{10}{3} = 3\frac{1}{3}$

FINDING EFFICIENT SOLUTIONS

Children should always be encouraged to adopt the most efficient methods whenever possible. For example, what is the best way to do the following:

$$\frac{5}{9} \times 63$$

We could attempt to do it the same way as above, turning the 63 into a top-heavy fraction...

$$\frac{5}{9} \times \frac{63}{1}$$

But this would lead to an unnecessary multiplication of 5 × 63 followed by a division by 9. Although correct, this is inefficient and unwieldy.

A much better approach is to remember the interchangeable use of '×' and the word 'of'.

So now we have:

$$\frac{5}{9} \text{ of } 63$$

In other words 'What is five ninths of 63?'

All we do now is follow the logic of finding one ninth of 63 by dividing 63 into 9 equal pieces (namely, 7). We then multiply this result by 5 to give us five ninths of 63. (And 5 × 7 = 35.)

This can be shown like:

$$5 \times \frac{63}{9}$$

$$5 \quad \times \quad 7 \quad = \quad 35$$

Answer: 35

Simple... and efficient!

Dividing a whole number by a fraction

This can often cause a lot of problems but really it needn't.

For example:

- $4 \div \frac{1}{2}$ means how many halves are there in the number 4.

Simply remembering the basics, we know there are 2 halves in **1 whole**, 4 halves in **2 whole**, 6 in **3 whole** and 8 in **4 whole**. Eight halves make the number 4.

$$4 \div \frac{1}{2} = 8$$

Alternatively we can turn the **whole** number into a top-heavy fraction and then divide the fractions, as shown in the box on page 327. (To convert any **whole** number into a fraction we simply divide that number by 1.)

Here is an example:

$$6 \div \frac{2}{3}$$

$$\rightarrow \quad \frac{6}{1} \div \frac{2}{3} \quad \rightarrow \quad \frac{6}{1} \times \frac{3}{2} \quad = \quad \frac{6 \times 3}{1 \times 2} \quad = \quad \frac{18}{2} \quad = \quad 9$$

DIVIDING A MIXED NUMBER BY A FRACTION

This is a classic type of question that causes endless difficulties:

- 'What is 2½ ÷ ¼?'

This is really quite straightforward.

We just need to remember that this question is the same as:

- 'How many quarters go into 2½?'

(Just as 15 ÷ 5 is the same as how many fives go into 15.)

To help you visualise the problem, simply picture 1 cake cut into 4 quarters. Now picture 2 cakes cut into quarters. So far we have eight quarters. Now we just need to consider how many quarters there are in one half of a cake. A whole cake has four quarters, so half a cake has two quarters.

In 2½ cakes there are a total of ten quarters.

Answer: 2½ ÷ ¼ = 10

Alternatively we can turn 2½ into an improper fraction (see page 314) and divide as described in the box on page 327, as follows:

$$\frac{5}{2} \div \frac{1}{4} \rightarrow \frac{5}{2} \times \frac{4}{1} = \frac{20}{2} = 10$$

Fractions and decimal numbers are directly related and fractions can be converted into their decimal form. **Converting fractions to decimal numbers** is much easier than children often think. This is especially true if you have a calculator! Bear in mind that in Year 5 and beyond children will have access to calculators.

- All the 'converting' examples below assume the use of a calculator and this will be common practice in the classroom.

For example, we know $\frac{5}{8}$ can be read as '*5 divided by 8*'. Simply typing this sum into our calculator gives us the decimal number equivalent to the fraction.

- $\frac{5}{8}$ = 5 ÷ 8 (*typed into calculator*) = 0.625

So if it's this easy to convert all fractions to their decimal number equivalent, why does it cause children problems? Maybe it just sounds difficult to '***convert fractions into their decimal form***'. Really, it is all about understanding the fraction in the first place, and knowing that $\frac{5}{8}$ can be read as '***5 divided by 8***'.

Here are some other examples:

$\frac{3}{5}$	=	3 ÷ 5 (*typed into calculator*)	=	0.6
$\frac{9}{10}$	=	9 ÷ 10 (*typed into calculator*)	=	0.9
$\frac{2}{7}$	=	2 ÷ 7 (*typed into calculator*)	=	0.285714285
$\frac{14}{4}$	=	14 ÷ 4 (*typed into calculator*)	=	3.5
$\frac{1}{8}$	=	1 ÷ 8 (*typed into calculator*)	=	0.125
$\frac{8}{8}$	=	8 ÷ 8 (*typed into calculator*)	=	1

A **very** common mistake is for children to think that the digits in the fraction must somehow reappear in the decimal number. For example, $\frac{1}{8}$ is often *incorrectly* written as **0.8** or **0.18**. If I could have a £1 for every time I'd seen this mistake, I'd... well, you know the rest.

One reason for this oh-so-common mistake ***might be*** because children are often introduced to decimal fractions (or decimal numbers) by way of tenths, hundredths and thousandths.

And

$\frac{1}{10}$ = 0.1	$\frac{1}{100}$ = 0.01
$\frac{2}{10}$ = 0.2	$\frac{2}{100}$ = 0.02
$\frac{3}{10}$ = 0.3	$\frac{3}{100}$ = 0.03 ...and so on.

The numerator ***does*** reappear in the decimal fraction when dealing with tenths, hundredths, thousandths...!

To convert **mixed numbers** to the decimal form, first convert to an improper fraction and then continue as above:

$2\frac{1}{4}$ = $\frac{9}{4}$ = 9 ÷ 4 (*typed into calculator*) = 2.25
$4\frac{3}{8}$ = $\frac{35}{8}$ = 35 ÷ 8 (*typed into calculator*) = 4.375

COMMON DECIMAL EQUIVALENTS

For some very common fractions, it will help your children to be efficient and confident if they just learn and know the decimal equivalent without having to work them out each time. These are:

$\frac{1}{2}$	=	0.5
$\frac{1}{4}$	=	0.25
$\frac{3}{4}$	=	0.75
$\frac{1}{10}$	=	0.1
$\frac{1}{100}$	=	0.01

And maybe also:

$\frac{1}{3}$	=	0.3333333... ('nought point three recurring'*)
$\frac{2}{3}$	=	0.6666666... ('nought point six recurring'*)
$\frac{1}{8}$	=	0.125
$\frac{3}{8}$	=	0.375
$\frac{1}{1000}$	=	0.001

*Recurring decimals are often denoted by putting a dot on top of the digit that repeats. If more than one digit repeats, a dot is put above the first and last repeating digits.

$\frac{1}{3}$	=	0.3333333... can be written as $0.\dot{3}$
$\frac{1}{7}$	=	0.142857142857142857... and can be written as $0.\dot{1}4285\dot{7}$

Percentages

Just as fractions are related to decimal numbers, so too are **percentages**.

The **'per cent'** symbol looks like this: **%**

'Per cent' literally means 'per hundred' but can be taken to mean 'out of 100' or 'divided by 100'.

So 75% is read as '*75 per cent*' and is the same as '75 out of 100' or '75 ÷ 100' or '75/100'.

Similarly 4% is read as '*4 per cent*' and is the same as '4 out of 100' or '4 ÷ 100' or '4/100'.

Converting percentages to decimal numbers is really easy. For example:

- 25% is the same as $\frac{25}{100}$ = 25 ÷ 100 = 0.25
- 50% is the same as $\frac{50}{100}$ = 50 ÷ 100 = 0.5
- 75% is the same as $\frac{75}{100}$ = 75 ÷ 100 = 0.75
- 100% is the same as $\frac{100}{100}$ = 100 ÷ 100 = 1

Likewise:

- 10% is the same as $\frac{10}{100}$ = 10 ÷ 100 = 0.1
- 20% is the same as $\frac{20}{100}$ = 20 ÷ 100 = 0.2
- 30% is the same as $\frac{30}{100}$ = 30 ÷ 100 = 0.3
- 40% is the same as $\frac{40}{100}$ = 40 ÷ 100 = 0.4

...and so on.

A very common mistake is to write 5% as 0.5, whereas in fact 5% *is the same as*

$$\frac{5}{100} = 5 \div 100 = 0.05$$

To convert decimal numbers to percentages, we need to do the above in reverse, and so instead of dividing by 100 we **multiply** by 100.

For example:

0.25	=	0.25 × 100 % *	=	25%
0.5	=	0.5 × 100 %	=	50%
0.75	=	0.75 × 100 %	=	75%
1	=	1 × 100 %	=	100%

* It is very important **always** to remember the 'per cent' symbol.
0.25 is **not** *the same as* 25, but it **is** *the same as* 25%.
Here the '%' symbol tells us it is: '*25 out of 100*'.

Converting fractions to percentages is just as straightforward.
First convert the fraction to its decimal equivalent and then do
as above.

For example:

$\frac{3}{10}$	=	3 ÷ 10	=	0.3	=	0.3 × 100%	=	30%
$\frac{1}{4}$	=	1 ÷ 4	=	0.25	=	0.25 × 100%	=	25%
$\frac{2}{5}$	=	2 ÷ 5	=	0.4	=	0.4 × 100%	=	40%

To convert percentages to fractions we need to replace the '%'
symbol with '*out of* 100' and just simplify the fraction (see page
320). For example:

70% is the same as '70 out of 100' $\quad = \frac{70}{100}$
Now we just simplify the fraction: $\quad \frac{70}{100} = \frac{7}{10}$

Another example:

15% is the same as '15 out of 100' $\quad = \frac{15}{100}$
Now we just simplify the fraction: $\quad \frac{15}{100} = \frac{3}{20}$

More examples:

26% is the same as '26 out of 100' $\quad = \frac{26}{100}$
Now we just simplify the fraction: $\quad \frac{26}{100} = \frac{13}{50}$

64% is the same as '64 out of 100' $\quad = \frac{64}{100}$
Now we just simplify the fraction: $\quad \frac{64}{100} = \frac{32}{50} = \frac{16}{25}$

> 20% is the same as '20 out of 100' $= {}^{20}\!/_{100}$
> Now we just simplify the fraction: ${}^{20}\!/_{100} = {}^{2}\!/_{10} = {}^{1}\!/_{5}$

A very common misconception is for children to think that 20% = ${}^{1}\!/_{20}$.

This probably stems from the fact that 10% *is* ${}^{1}\!/_{10}$. This may then lead to the false assumption that the percentage amount must reappear as the denominator of the fraction. We need to regularly remind children what the percentage symbol *means*!

It is worth just knowing these following few, without having to work them out each time:

25%	=	¼	(one quarter)
50%	=	½	(one half)
75%	=	¾	(three quarters)
100%	=	1	(one whole)
10%	=	${}^{1}\!/_{10}$	(one tenth)
1%	=	${}^{1}\!/_{100}$	(one hundredth)

So fractions, decimal numbers and percentages are all related. It is really important for children to *know* this.

Number Lines can be used to illustrate the relationships:

0	0.25	0.5	0.75	1

0	¼	½	¾	1
0	25%	50%	75%	100%

0	0.1	0.2	0.3	0.4	0.5	0.6	0.7	0.8	0.9	1

0	${}^{1}\!/_{10}$	${}^{2}\!/_{10}$	${}^{3}\!/_{10}$	${}^{4}\!/_{10}$	${}^{5}\!/_{10}$	${}^{6}\!/_{10}$	${}^{7}\!/_{10}$	${}^{8}\!/_{10}$	${}^{9}\!/_{10}$	1
0	10%	20%	30%	40%	50%	60%	70%	80%	90%	100%

We often need to **find a percentage of a number**. This is a really worthwhile life skill and one that can be easily accomplished without a calculator using just a pencil and paper. In my experience the percentage (%) button on a calculator is often more of a hindrance than a help. It is often not at all clear, to children or adults, how the function works, or which number you are calculating the percentage from. I find it best to follow the method set out below.

Children will be asked to find simple percentages of whole number quantities, *without* using a calculator. For example:

- 'What is 20% of 400?'
- 'Find:
    30% of £60,
    5% of 220cm,
    17.5% of 80g,
    11% of £300'
- 'A netball team played 20 games. They won 65% of their games. How many games did they win?'
- 'In a sale a pair of jeans is reduced by 25%. They originally cost £50. What do they cost now?'

All of these questions, and ones like them, can be dealt with using a simple method of recording stages in a 'table'. But before we begin, it might be helpful to remember that the whole '*thing*', whatever that might be – the whole amount, the whole measurement, the total number of games, the original cost, the full price, the total bill – **is equivalent to 100%**.

Find:
30% of £60	5% of 220cm	17.5% of 80g	11% of £300
↑	↑	↑	↑
This is the 100%	This is the 100%	This is the 100%	This is the 100%

Finding 10 per cent (10%) of a number is usually simple, and this is our standard starting point. To find 10% of any number simply divide that number by 10. Why? Because 100 per cent (100%) divided into 10 equal pieces is 10 per cent (10%), and once you know what 10% of a number is, the rest is easy:

- To find 20%, simply double your answer for 10%.
- To find 5%, halve your answer for 10%.
- To find 2.5%, halve your answer for 5%.
- To find 30%, add your answer for 10% to your answer for 20%.
- To find 35%, add your answer for 30% to your answer for 5%.
 …and so on.

A 'table' helps you keep track of your workings as you build up your answers.

To answer: 'What is 20% of 400?' we can draw out a table like the one below:

Percentage		Number
100%	is	400

Start by finding 10% of 400. Divide 400 by 10.

10%	40

Now find 20% by doubling. Double 40.

20%	80

So the **answer** is: 80

Similarly for the following:

Find: 30% of £60

Percentage		Number
100%	is	60

Start by finding 10% of 60. Divide 60 by 10.

10%	6

Now find 20% by doubling. Double 6.

20%	12

Now add your 'answer' for 10% to your 'answer' for 20%. (6 + 12)

30%	18

Answer: £18

Find: 5% of 220cm

Percentage		Number
100%	is	220

Start by finding 10% of 220. Divide 220 by 10.

10%	22

Now find 5% by halving 22.

5%	11

Answer: 11cm

Find: 17.5% of 80g

Percentage		Number
100%	is	80

Start by finding 10% of 80. Divide 80 by 10.

10%	8

Now find 5% by halving. Halve 8.

5%	4

Now find 2.5% by halving. Halve 4.

2.5%	2

Now add your 'answers' for 10%, 5% and 2.5% together. (8 + 4 + 2)

17.5%	14

Answer: 14g

Find: 11% of £300

Percentage		Number
100%	is	300

Start by finding 10% of 300. Divide 300 by 10.

10%		30

Now find 1% by dividing the 10% by 10. Divide 30 by 10.

1%		3

Now add your 'answers' for 10% and 1% together. (30 + 3).

11%		33

Answer: £33

Finding 10% is always a good starting point. Some children will be really happy with this 'method' and won't want to deviate from this common approach. That's fine; others may be happy to explore different starting points.

In the next example it may be more efficient to start by finding 50%. This is easy as 50% is exactly one half of 100%.

A netball team played 20 games. They won 65% of their games. How many games did they win?

Percentage		Number
100%	is	20

Start by finding 50% of 20. Halve 20.

50%		10

Now find 10%. Divide 20 by 10.

10%		2

Now find 5% by halving 10%. Halve 2.

5%		1

Now add your 'answers' for 50%, 10% and 5% together. (10 + 2 + 1)

65%		13

Answer: 13 games

● But if your children are happier to start by finding 10% then…

Percentage		Number
100%	is	20

Start by finding 10%.

10%	2

Now find 20%.

20%	4
30%	6
40%	8
50%	10
60%	12
5%	1

Now add your 'answers' for 60% and 5%. (12 + 1)

65%	13

Answer: 13 games

The next question has two stages to it. Often pupils will forget to do the second stage and therefore not answer the question correctly.

• 'In a sale a pair of jeans is reduced by 25%. They originally cost £50. What do they cost now?'

We would start this question exactly as we have done with the ones above. We'd find 25% of £50, using the table to help record our 'workings'. We could start by finding 10%, as with most of the examples above. Alternatively we could find 50% and then halve this to find 25%.

Both are shown below:

Percentage		Number
100%	is	50

Start by finding 10% of 50. Divide 50 by 10.

10%	5

Now find 20% by doubling. Double 5.

20%	10

Now find 5% by halving 10%. Halve 5.

5%	2.5

Now add your 'answer' for 20% to your 'answer' for 5%. (10 + 2.5)

25%	12.5

Or:

Percentage		Number
100%	is	50

Start by finding 50% of 50. Halve 50.

50%	25

Now find 25% by halving. Halve 25.

25%	12.5

So the reduction in the sale is: £12.50

Now for the second stage of the question (which lots of pupils will 'forget' to do):

What do they cost now?
The jeans now cost £12.50 less. So that is £50 − £12.50 = £37.50
Answer: £37.50

Alternatively, this question can be answered in one stage. *Some* children may realise that if the jeans are reduced by 25% they must now cost 75% of the original price. By finding 75% they will arrive at the final answer. This is a far more efficient approach and should be encouraged if children are comfortable with it.

In some cases, where the numbers aren't quite so simple, using a calculator can be helpful – but the method remains exactly the same, and **doesn't** involve the use of the '%' button. Questions children might be expected to answer – with the help of a calculator – are:

Find:
- 30% of £263
- 75% of 670cm
- 12% of £782.

Again we start with a 'table'. We use the calculator only to help do some of the division – now the numbers aren't quite so straightforward.

⬡ 30% of £263:

Percentage		Number
100%	is	263

Start by finding 10% of 263. Divide 263 by 10, using a calculator if needed.

10%		26.3

Now find 20% by doubling. Double 26.3, using a calculator if needed.

20%		52.6

Now add your 'answer' for 10% to your 'answer' for 20%. (26.3 + 52.6 = 78.9)

30%	=	78.9

Answer: £78.90

⬡ 75% of 670cm:

Percentage		Number
100%	is	670

Start by finding 50% of 670. Halve 670, using a calculator if needed.

50%		335

Now find 25% by halving. Halve 335, using a calculator if needed.

   25%   167.5

Now find 75% by adding together your 'answers' for 50% and 25%.

   75%   =   502.5

Answer: 502.5cm

12% of £782:

Percentage		Number
100%	is	782

Start by finding 10% of 782. Divide 782 by 10, using a calculator if needed.

   10%   78.2

Now find 1% by dividing the 10% by 10. Divide 78.2 by 10.

   1%   7.82

Now find 2% by doubling 1%. Double 7.82.

   2%   15.64

Now add your 'answer' for 10% to your 'answer' for 2%. (78.2 + 15.64 = 93.84)

   So 12%   =   93.84

Answer: £93.84

Percentages are taught in different ways, in different classrooms, in different schools. However, children usually like to have a standard starting point. It makes them feel comfortable and confident, armed with a method they can always fall back on. Using a 'table' like the one in the examples above, gives them a familiar and consistent approach. It is logical and reliable, neat and efficient. Most importantly, children like it. I haven't seen a better 'starter method'.

That said, an appropriate future goal for many pupils will be to appreciate the efficiency of finding percentages by using multiplication alone. Picking an example from those above:

🎲 Find 75% of 670cm.
As 75% is the same as 0.75 and the word 'of' can be interchanged with '×' we have: 0.75 × 670, which = 502.5
Answer: 502.5cm

Similarly, if a price increases by **10%**, it is important for children to realise the new amount is now **110%** or the original. (**110%** is equivalent to the decimal number **1.1**)

🎲 Petrol prices increase by 10%. Before the price increase petrol cost £1.20 a litre. What does it cost now?
Answer: 1.1 × 1.20 = 1.32   Petrol now costs £1.32 per litre.

Likewise, if a price decreases by **10%**, it is important for children to realise the new amount is now **90%** of the original. (And **90%** is equivalent to the decimal number **0.9**)

🎲 A pair of jeans originally cost £70. They are then reduced by 10% in the sale. What is the new cost?
Answer: 0.9 × 70 = 63   The jeans now cost £63.

There are other types of questions involving percentages that ask us to find more than just a percentage as an answer. Children will be asked questions along the following lines:

- 'Calculate as percentages: 18 out of 32 and 24 out of 60'
- 'Joey got 45 out of 60 in a maths test. What percentage is this?'
- 'Megan got 30 out of 60 in a spellings test. Sophie got 45%. Who did best?'
- '42% of the children in a class are girls. What percentage are boys?'

- '85% of children have school lunches. What percentage of children do not?'
- '12 raisins are 20% of all the raisins in this little box. How many raisins altogether are in the box?'

As we've seen in other chapters and earlier in this one, the key thing is to interpret the question correctly. Time spent ensuring that your children are getting to grips with *exactly* what they are being asked is very well spent indeed.

The first three examples above all rely on knowing how to convert fractions to percentages (see page 336). First, convert the fraction to its decimal equivalent (by dividing the numerator by the denominator using a calculator) and then convert to a percentage by multiplying by 100%.

Calculate as percentages: 18 out of 32 and 24 out of 60
Answers:
18 out of 32 = 18/32 = 0.5625 = 56.25%
24 out of 60 = 24/60 = 0.4 = 40%

Joey got 45 out of 60 in a maths test. What percentage is this?
45 out of 60 = 45/60 = 0.75 = 75%
Answer: 75%

Megan got 30 out of 60 in a spellings test. Sophie got 45%. Who did best?
30 out of 60 = 30/60 = 0.5 = 50%
50% is a higher score than 45%.
Answer: Megan did best.

The next two questions simply rely on the knowledge that the whole amount is equivalent to 100%.

42% of the children in a class are girls. What percentage are boys?
100% − 42% = 58%
Answer: 58% of the class are boys.

85% of children have school lunches. What percentage of children do not?
100% − 85% = 15%
Answer: 15% of children do not have school lunches.

The final question still relies on the knowledge that the whole amount is 100%, but is a little more complex so a table similar to the ones we have seen previously will help a lot.

---

12 raisins are 20% of all the raisins in this little box. How many raisins altogether are in the box?

Percentage		Number
100%	is	? (We don't know this yet. This is what we need to find out.)

Start by writing down what we do know:

20%		12

Now find 10% by halving. Halve 12.

10%		6

Now find 100% by multiplying by 10:

100%	is	60

Answer: There are 60 raisins altogether in the box.

Ratio and proportion

Some time from Year 6 onwards children will be introduced to **ratio** and **proportion**.

A ratio is a way of comparing two quantities and a proportion is a way of comparing a quantity to the whole amount. Both ratios and proportions are expressed using fractions and percentages.

The difference between proportion and ratio can be summed up as: A **proportion** compares a 'part' to the whole. A **ratio** compares a 'part' to a 'part'.

I will give you an example of each to show you what I mean.

Proportion

At Matthew's birthday party there are 4 girls and 12 boys. So there is a total (or whole amount) of 16 children. The proportion of girls at the party is 4 out of 16. This can also be expressed as $\frac{4}{16}$ or simplified to ¼ or '1 in 4' or 25% (see also page 337).

Similarly, the proportion of boys at the party is 12 out of 16, which can also be expressed as $\frac{12}{16}$ or ¾ or '3 in 4' or 75%. So, to find the **proportion**, each 'part' – girls or boys – is compared with the whole amount.

Ratio

At Matthew's birthday party there are still 4 girls and 12 boys.

The **ratio** of *girls to boys* is '4 to 12' and this is written using ratio notation as 4:12.

Ratios can be simplified in much the same way as fractions, by dividing each 'part' by the same number. So, in this case, dividing both 'parts' by 4, the equivalent ratio of girls to boys is 1:3. This means for every 1 girl, there are 3 boys.

The **ratio** of *boys to girls* is '12 to 4' and using the ratio notation this is written as 12:4.

Cancelling or simplifying gives the equivalent ratio for boys to girls as 3:1. This means for every 3 boys, there is 1 girl – or alternatively – there are 3 times as many boys as girls.

So for ratio, each 'part' – girls or boys – is compared with the other 'part'.

Understanding the connection between ratio and proportion means that given only the **ratio** we can deduce the **proportion**. Given only the **proportion** we can deduce the **ratio**.

The **ratio** of *girls to boys* at Matthew's party was 4:12.

To find the **proportion** of girls is simple.

We simply need to add together all the 'parts' of the ratio to find the total.

$$4 + 12 = 16.$$

The **proportion** of girls is therefore '4 out of 16' or $\frac{4}{16}$ which can be simplified to $\frac{1}{4}$.

On the other hand if we only knew the **proportion** of girls at Matthew's party we could just as easily find the **ratio**.

The **proportion** of girls is $\frac{4}{16}$.

The **proportion** of boys must therefore be $\frac{12}{16}$ (because if there are 4 girls out of 16, the rest must be boys). The ratio of *girls to boys* is therefore 4:12, which can be simplified to 1:3.

Mia likes to make a particular shade of purple paint using red, blue and white paint in the ratio of 3:4:1.

(That is, for every 3 parts of red paint she uses 4 parts of blue and 1 part of white.)

What proportion of blue paint does Mia use?

The total number of 'parts' is $3 + 4 + 1 = 8$.

Mia uses 4 'parts' of blue paint.

The proportion of blue paint is '4 out of 8' or 4/8, which can be simplified to $\frac{1}{2}$.

Answer: $\frac{1}{2}$

The proportion of the class learning to play the recorder is $\frac{12}{30}$.

What is the ratio of those learning the recorder to those not learning the recorder?

The proportion of the class learning to play the recorder is $\frac{12}{30}$, so the proportion of the class not learning to play the recorder must be $\frac{18}{30}$ (because 12 and 18 make 30, and there are only two options: 'to learn' or 'not to learn').

The ratio of 'learning' to 'not learning' is therefore 12:18, which can be simplified (by dividing each 'part' by 6) to 2:3.

Answer: 2:3

Ratio is typically used to solve maths problems. Here are some very common examples:

⬡ Mindu spends her £60 birthday money on DVDs and clothes in the ratio of 1:4. How much does she spend on clothes?

DVDs		Clothes
1	:	4

There are 5 'parts' in total   (1+ 4 = 5)
£60 divided by 5 equals £12   (60 ÷ 5 = 12)
Each 'part' is therefore equal to £12
4 'parts' is therefore equal to 4 × £12   (4 × 12 = 48)

DVDs		Clothes
1	:	4
£12	:	£48

Answer: Mindu spends £48 on clothes.

⬡ There are 36 pupils in this Year 6 class. The ratio of boys to girls is 5:4. How many boys are in the class?

Boys		Girls
5	:	4

There are 9 'parts' in total   (5 + 4 = 9)
36 pupils divided by 9 equals 4   (36 ÷ 9 = 4)
Each 'part' is therefore equal to 4 pupils
5 'parts' are therefore equal to 5 × 4 pupils (5 × 4 = 20)

Boys		Girls
5	:	4
20 pupils	:	16 pupils

Answer: There are 20 boys in the class.

⬡ A fruit drink is made by mixing juice and water in the ratio of 1:9. Isaac wants to make 200 millilitres of drink. How much juice and how much water should he use?

Juice		Water
1	:	9

There are 10 'parts' in total   (1+ 9 = 10)
200 millilitres divided by 10 equals 20 ml   (200 ÷ 10 = 20)

Each 'part' is therefore equal to 20 ml
9 'parts' is therefore equal to 9 × 20 ml   (9 × 20 = 180)

Juice		Water
1	:	9
20 ml	:	180 ml

Answer: Isaac needs 20 millilitres of juice and 180 millilitres of water.

The angles in a triangle are in the ratio of 1:2:3. Find the size of each angle.

To solve this problem we must first know that the angles in a triangle always add up to 180 degrees (°).

Angle 1		Angle 2		Angle 3
1	:	2	:	3

There are 6 'parts' in total   (1 + 2 + 3 = 6)
180° divided by 6 equals 30°   (180 ÷ 6 = 30)
Each 'part' is therefore equal to 30°
2 'parts' are therefore equal to 2 × 30°   (2 × 30° = 60°)
3 'parts' are therefore equal to 3 × 30°   (3 × 30° = 90°)

Angle 1		Angle 2		Angle 3
1	:	2	:	3
30°	:	60°	:	90°

Answer: The 3 angles are 30°, 60° and 90°.

Direct proportion is something pupils will be expected to use and is something that I think sounds harder than it actually is.

Here is a simple example to demonstrate direct proportion:

---

2 apples cost 40p. How much do 4 apples cost?
Quite simply the number of apples has doubled and so too must the cost.
Answer: 80p

And there you have it – direct proportion: as one quantity changes so too must the other, by the same ratio.

🎲 How much do 6 apples cost?
The number of apples has now trebled, so too must the cost.
Answer: 120p

We most often use direct proportion when calculating quantities for a recipe. We possibly do this without thinking and most probably without realising the involvement of 'direct proportion'!
For example:

🎲 **Cottage pie for 2**
    200g minced beef
    80g carrots
    500g mashed potato
If we wanted to make this recipe for 4 people, we would simply double all the quantities.
Answer: 400g minced beef, 160g carrots and 1000g (that is 1kg) mashed potato.

But if we wanted to make this recipe for 3 people, or 5 people, or 9 people, or… we might need to think a little more carefully.

The most common technique is to calculate the recipe for 1 person (the 'unit' method). Then multiply each quantity by the number of people.

So the above 'recipe for one' is:

- 100g minced beef, 40g carrots and 250g mashed potato.

And for 5 people?

It is now simple. We just multiply each quantity in the 'recipe for one' by 5:

- 500g minced beef, 200g carrots and 1250g mashed potato.

I do hope this chapter has reduced any anxieties you and your child had about fractions.

CONCLUSION

Thank you so much for investing time and energy to read this book. I really hope it has helped you and your children – and that you will continue to find it helpful in the years to come.

One of the most common comments from parents of secondary age pupils is '...*but my child did this at primary school*'. The important point about maths is that concepts have to be revisited and practised many times. Just because children have been introduced to a concept and shown the mechanics, it doesn't mean they are fluent with the idea.

Think of it as learning to drive a car. Someone may show you how the clutch/accelerator/ brake/gears work – and you may even have a 'go' with that person sitting beside you. With much practice and guidance you may even have the car going in the right direction and at a reasonable speed. But left alone too early and you might crash. Before you are allowed out in the car alone you must practise, practise, practise. And even then in unfamiliar surroundings or with busy traffic or bad weather you will have to concentrate really hard to remember everything you have been taught.

This is exactly the same for children learning maths. They may have been shown how to do long multiplication – but left alone too early and the calculation will undoubtedly crash. Often, as with long multiplication, there is a lot of stuff going on at once – interwoven strands of mathematical understanding (for example, counting, adding, sequencing, understanding of place value...) and children need to concentrate hard. Practice, practice and more practice is required – before the child can truly go solo. And this is what is sometimes so difficult to express to parents: maths takes time!

Have confidence in your children. Keep practising, keep encouraging, keep supporting. Remember the importance and don't give up. Don't opt for the easy option and claim your children are 'no good at maths'. It is way too early to say at 7, 11 or 13 that your children are not cut out for maths (or rugby, swimming, or anything else in fact). Keep all the doors wide open.

And remember, with maths there is not always just one answer. Mathematicians are continually pushing boundaries, experimenting and questioning. Mathematics is not now, nor ever has been, a fixed subject. As fast as the human mind can conjure up new questions, so the field of mathematics evolves to find ways to solve them.

Too many children in too many classrooms are led to believe that mathematics is fixed and finished, a rusty old relic of a subject only concerned with abstract rules and formulae.

They miss out on the fun – the playing, the trying, the experimenting, the discovering, the seeing it for themselves.

Yes, there are basic rules to follow. But, just as with a language, once you know those basic rules you can have fun creating and developing ideas.

I wish your children a mathematical journey full of fun, confidence and discovery.

GLOSSARY

There has simply not been the capacity to include all primary maths topics in detail in this one book. For the three omitted topics (Shape; Measuring; and Statistics) key terms, definitions, rules and formulae have been included in the following glossary. The terms are *not* alphabetical. Instead they are listed in an order that makes most sense while reading.

Shapes

2-D (two dimensional) **shapes**: these are shapes with just two dimensions – length and width. They are flat shapes that we can draw on paper, such as a circle, triangle, square or rectangle.

3-D (three dimensional) **shapes**: these are shapes with three dimensions – length, width *and* height. They are solid shapes that we can imagine holding, such as a sphere (*think of a football*), cone (*think of an ice-cream*), cube (*think of an Oxo cube*), cuboid (*think of a shoe box*), cylinder (*think of a baked bean tin*) or pyramid (*think of the Pyramids in Egypt*).

Net: what a 3-D shape would look like if it was opened out flat.

Polygons: 2-D shapes with straight lines that join together. Regular polygons have sides of equal length and angles of equal size. Irregular polygons don't! See Appendix (page 363) for a table of common polygons.

Polyhedron: 3-D shapes that have **faces** (flat sides/surfaces), **edges** (the sharp bit you can run your finger along) and **vertices** (corners or points). See Appendix (page 364) for a table of common polyhedrons.

Equilateral triangle: a triangle with 3 equal sides.

Isosceles triangle: a triangle with 2 equal sides.

Scalene triangle: a triangle with no equal sides.

Right-angled triangle: a triangle that has a right angle (90 degrees).

Quadrilateral: a shape with 4 straight sides and 4 angles. Common quadrilaterals are rectangle, square, parallelogram, rhombus, trapezium and kite.

Angle: a measure of turn.

Right angle: a quarter turn or 90 degrees.

Acute angle: an angle less than 90°.

Obtuse angle: an angle greater than 90° but less than 180°.

Reflex angle: an angle greater than 180°.

Straight line: an angle of exactly 180° (which is the same as 2 quarter turns or 2 right angles).

A whole turn: an angle of 360°.

Set square: a piece of equipment used to draw right angles accurately.

Protractor: a piece of equipment used to measure angles.

Angle-measurer: a piece of equipment used to measure angles, often circular.

Vertically opposite angles: angles formed when two straight lines cross; the angles opposite each other are always the same.

Internal angle: the angle on the inside of a polygon formed by 2 of its sides. If a shape is regular, all the internal angles will be identical.

External angle: the angle on the outside of a polygon formed by one of its sides and a line extended from an adjacent side. A good way to picture this is to imagine the angle through which you need to turn to change direction – from one side to the next. (Note: an adjacent exterior and interior angle will always add up to 180° because they form a straight line.)

Parallel lines: straight lines that are the same distance apart all along their lengths.

Perpendicular: two lines are perpendicular to each other if one meets the other at right angles.

Intersect: the point at which two straight lines touch or cross each other.

Bisector: a line that cuts another line or angle exactly in half.

Locus: a pathway of a point, which moves according to a certain rule. A very common locus is the drawing of a circle, the rule being that each point along the way is exactly the same distance from the centre. The plural for locus is **loci**.

Circumference: a line drawn to create a circle.

Diameter: the straight-line distance across a circle passing through the centre of the circle.

Radius: the straight-line distance from the centre of the circle to any point on the circumference. The radius is always exactly half the length of the diameter. The plural for radius is **radii**.

Chord: a straight line that joins any two points on the circumference. The diameter is a chord that passes through the centre of the circle and is the longest possible chord of a circle.

Arc: a curved line. A section of a circumference is an arc.

Sector: the portion of a circle 'made' by two radii and an arc. Imagine a slice of pie.

Segment: the portion of a circle 'made' by a chord and an arc. Imagine a segment of orange.

Concentric circles: circles inside circles all with the same centre.

Semi-circle: exactly half of a circle.

Co-ordinates: the position of a point on a graph or grid. Co-ordinates are shown as a pair of numbers within brackets. The first number is the x-co-ordinate, the second the y-co-ordinate.

The origin: the starting point on a graph or grid, given the co-ordinate (0, 0). To find the co-ordinate (1, 3), for example, we start at the origin and move across 1 square and up 3.

x-axis: the horizontal axis on a graph. (Think: 'a-cross'!)

y-axis: the vertical axis on a graph.

Transformation: a change in the shape, size or position of an object. Examples include reflection, translation, rotation, enlargement, as defined below.

    Reflection: this transforms a shape into its mirror image.

    Translation: this transforms a shape by moving its position – up or down, or from side to side. The object itself remains unchanged (that is, its shape, size and orientation are still the same).

    Rotation: transforms a shape by turning it around a specified point.

    Enlargement: transforms a shape by making it bigger or – rather bizarrely – smaller.

Symmetry: when one half of a shape is a mirror image of the other. A shape is said to be symmetrical if it can be 'cut' in half so that each half is an exact reflection of the other.

Congruent: this describes objects that are identical in size and shape.

Perimeter: the distance around the outside of a shape.

Area: the amount of space that a shape covers. We measure all areas – big and small – in squares.

Compound shapes: 2-D shapes that are made up from other shapes.

Surface area (of a 3-D shape): the total area of all the faces.

Measuring

Standard units: universally agreed units of measure such as centimetres, kilograms and litres.

Non-standard units: units of our choosing that we might use to make comparisons. Examples of these could be our hand span or the length of our stride.

Length: how long something is. Words associated with length are: distance, long, short, tall, high, low, wide, narrow, deep, shallow, thick, thin, far, near, close. (See Appendix page 365.)

Mass: how heavy something is. Words associated with mass are: weight, weighs, heavy, light, balances. (See Appendix page 366.)

Capacity: the amount that a container can hold. Words associated with capacity include: full, empty, holds. (See Appendix page 367.)

Volume: how much space something takes up. (See Appendix page 367.)

Time: how long something takes to occur, measured in seconds, minutes, hours, days, weeks, months, years, decades, centuries and so on. (See Appendix page 368.)

Statistics

Data: another word for information (singular **Datum**).

Handling data: refers to collecting, sorting and organising data; drawing tables, graphs and charts; extracting and interpreting data in tables, graphs and charts; and estimating likely probabilities. Otherwise known as **Statistics**.

Frequency: how often something occurs. As a mathematical term it means much the same as its everyday usage.

Table: record of results or information.

Tally marks: marks made in bundles of five: four straight lines with a diagonal crossed through to make the fifth mark. They are used to keep a count (or 'tally').

Frequency table: this is created by simply adding up the tally marks in each category.

Averages: mean, median, mode: as defined below.

Mean: probably the most commonly known average and it is often the one people think of when they hear the word 'average'. To calculate the mean you simply add together all the numbers and then divide this total by however many numbers you had. Can only be used for numerical data. Sportspeople know it as the 'batsman's average'.

Median: the middle bit of data when (and only when) all the data have been listed in size (rank) order from smallest to largest. Can only be used for numerical data.

Mode: the most frequently occurring 'thing' in the list. Can be used for numerical data and non-numerical data.

Range: used to measure how spread out (or dispersed) the data are. Can only be used for numerical data. To calculate the **range** you simply subtract the smallest data value from the largest.

Graphs, charts and diagrams: all ways of displaying data pictorially.

Bar chart (or bar graph): this a graph that illustrates the frequency of each category by the height of the bar.

Vertical line diagram (or bar line chart or bar line graph): this is similar to a bar chart except that the width of each bar is just the width of a line.

Pictogram: a diagram with pictures used to illustrate data. The more pictures, the more data.

Pie chart: another type of picture or graph used to illustrate data. Pie charts are based on simple circles. The whole pie (or circle) is 'cut' into slices (or sectors), with each slice representing a different proportion of the whole. The bigger the slice, the more data in that category.

Construct: as a mathematical term, this simply means 'to draw'. Pupils may be asked to construct a tally chart, a bar chart, pictogram, line graph, pie chart or any other diagram.

Analyse: to interpret data and reach conclusions.

Venn diagrams and **Carroll diagrams:** these are used to sort and display information.

Qualitative data: these are non-numerical (non-number) data. Examples of qualitative data include: favourite colours, foods, television programmes, ways of getting to school, voting preference and so on.

Quantitative data: these are numerical (number) data. Examples of quantitative data include: shoe size, number of brothers and sisters, age, height, property values, population numbers, unemployment figures and so on. Quantitative data are either discrete or continuous, as defined below.

Discrete data: these have precise, exact, specific values. Examples include: test scores, shoe sizes, number of pets, number of brother and sisters, number of words on a page, number of peas in a pod and so on.

Continuous data: these make up the rest of the numerical data and are often associated with measurement. Examples of continuous data include: height, weight, time and so on.

Questionnaires and **surveys:** these are common ways to collect data.

Database: somewhere to store the collected data, such as a spreadsheet.

Hypothesis: this is a simple statement, theory or idea that forms the basis for a test. Once the test has been performed, the hypothesis can be confirmed as either 'True' or 'False'.

Prediction: something that we believe may be true but which needs to be tested further.

Distribution: summary of the data values and their associated frequencies.

Normal distribution: information displayed as a graph that has a bell-like shape. It is called 'normal' because in all kinds of data, it is very common for the middling values in a range to have the highest frequency, and the outliers the lowest.

Probability: a measure of how likely (or unlikely) something is to happen. All probabilities can be placed on a scale from 0 to 1, where 0 means impossible and 1 represents certainty.

APPENDIX

COMMON POLYGONS

Triangle – has 3 sides and 3 angles

Quadrilateral – has 4 sides and 4 angles

Pentagon – has 5 sides and 5 angles

Hexagon – has 6 sides and 6 angles

Heptagon (or Septagon) – has 7 sides and 7 angles

Octagon – has 8 sides and 8 angles *(think octopus)*

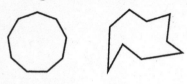

Nonagon – has 9 sides and 9 angles

Decagon – has 10 sides and 10 angles *(think decade)*

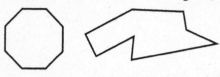

COMMON POLYHEDRONS

Cube

6 faces
12 edges
8 vertices

Cuboid

6 faces
12 edges
8 vertices

Square based pyramid

5 faces
8 edges
5 vertices

Tetrahedron
(Triangular based pyramid)

4 faces
6 edges
4 vertices

Triangular prism

5 faces
9 edges
6 vertices

Note: Spheres, hemispheres, cones and cylinders are not polyhedrons as their faces are not flat surfaces (polygons).

LENGTH

We usually use millimetres (mm), centimetres (cm), metres (m) or kilometres (km) for measuring lengths. These are called **metric units**.

1 centimetre (cm)	=	10 millimetres (mm)
1 metre (m)	=	100 centimetres (cm)
1 kilometre (km)	=	1000 metres (m)

Sometimes the old **imperial units** of length are still used: inches, feet, yards and miles.

1 foot	=	12 inches
1 yard	=	3 feet
1 mile	=	1760 yards

Because of our historic use of imperial units, some are still in common use in the UK (and widely used in the USA), especially the 'mile'. So it is handy to know rough conversions between the two types of units:

1 inch	is roughly	2.5cm
1 foot	is roughly	30 cm
1 yard	is roughly	1 m
1 mile	is roughly	1.5 km

(5 miles = 8 km is a handy and pretty accurate conversion to know.)

MASS

Mass is measured in the **metric units** of grams (g), kilograms (kg), and tonnes.

1 kilogram (kg)	=	1000 grams (g)
1 tonne	=	1000 kilograms (kg)

Most of us probably measure 'mass' or weight less frequently than we measure 'lengths'.

For this reason children tend to have much less of an idea of what things might weigh. So it is a good idea to give children a couple of examples for easy reference and comparison to help them to estimate:

100g	is roughly the weight of a small apple
1kg	is the weight of a typical bag of sugar

The old **imperial units** for mass are: ounces, pounds, stones and tons.

```
16 ounces   =   1 pound
1 stone     =   14 pounds
```

These imperial units are rarely used in maths lessons now but it can still be handy to know a rough conversion between the two types of units, especially for recipes:

```
1 ounce    is roughly   25g
1 pound    is roughly   0.5 kg (or 454g to be accurate)
1 kg       is roughly   2.2 pounds
```

CAPACITY AND VOLUME

Capacity and volume both measure how much space something takes up.

The terms 'capacity' and 'volume' are often used interchangeably. They do measure much the same thing, although they can be distinguished: capacity is a measure of how much substance a container can hold; volume is a measure of the amount of space the container occupies. So capacity describes the amount of stuff that can go into a container, while volume describes the amount of space taken up by the container. It's a fine distinction, I know, which is why 'capacity' and 'volume' are often synonymous in practice.

Capacity is measured in the **metric units** of millilitres, centilitres and litres.

```
1 centilitre (cl)  =  10 millilitres (ml)
1 litre (l)        =  100 centilitres (cl)  =  1000 millilitres(ml)
```

When estimating capacity it is useful to have some examples for reference:

```
5ml        is roughly the amount of liquid a typical teaspoon
           can hold (and is the exact measure of a typical
           medicine spoon)
```

I litre   is equal to a typical carton of juice
            from the supermarket

The old **imperial units** for capacity are: fluid ounces, pints and gallons.

20 fluid ounces	=	1 pint
1 gallon	=	8 pints

These imperial units are rarely used in maths lessons now but it can still be handy to know a rough conversion between the two types of units:

1 pint	is roughly	0.5 litre
1 gallon	is roughly	4.5 litres

Volume is measured in cubic units such as cubic millimetres, cubic centimetres and cubic metres.

One cubic centimetre is simply a cube with each side being one centimetre long:

To measure volume, we need to consider how many of these cubes would fit into the three-dimensional shape. Children often start measuring the volume of a container by seeing how many 1-centimetre multilink cubes can fit into it.

1 cubic centimetre	= 1000 cubic millimetres
1 cubic metre	= 1 000 000 cubic centimetres

TIME

1 minute	=	60 seconds
1 hour	=	60 minutes
1 day	=	24 hours

```
I week       =  7 days
I fortnight  =  14 days
I year       =  12 months  =  52 weeks*  =  365 days
I leap year  =  366 days
```

* 365 days are not exactly divisible by 7, so a year is actually 52 weeks and I day – which is why your birthday moves on to the next day of the week each year. Or two days on if it's leap year and your birthday is after February…

```
I month      =  28, 29, 30 or 31 days
```

There is a rhyme to remember the number of days in each month:

'Remember 30 days has September, April, June and November.
All the rest have 31, except for February alone,
which has 28 days clear but 29 each leap year.'

For me this never really helped as I could never remember the rhyme!

Instead I have always used my knuckles! If this is new to you then this is how it works.

Make a fist out of both hands and put them side by side so that all the knuckles are pointing upwards. Squeeze your hands together so that there is no gap (or dip) between the two fists. A knuckle represents 31 days, a dip between the knuckles represents 30 days. Start with January and read from left to right:

January	-	knuckle	-	31 days
February	-	dip	-	*28 days (or 29 in a leap year)
March	-	knuckle	-	31 days
April	-	dip	-	30 days
May	-	knuckle	-	31 days
June	-	dip	-	30 days
July	-	knuckle	-	31 days
August	-	knuckle	-	**31 days
September	-	dip	-	30 days

October   -   knuckle   -   31 days
November -   dip   -   30 days
December -   knuckle   -   31 days

* February is in a dip, but is the exception to the rule as it has
only 28 or 29 days.
** There is no dip between the two hands so both July and
August are 'knuckles' and have 31 days.

1 decade	=	10 years
1 century	=	100 years
1 millennium	=	1000 years

To convert from the 12-hour clock with am/pm to the 24-hour clock

Between:

12 midnight	1.00 am	1.00 pm
to 12.59 am	to 12.59 pm	to 11.59pm
Subtract 12 hours	**Straight conversion**	**Add 12 hours**

Examples:

12 midnight → 0000 hours	1.00 am → 0100 hours	1.00 pm → 1300 hours
12.04 am → 0004 hours	2.30 am → 0230 hours	4.30 pm → 1630 hours
12.15 am → 0015 hours	7.00 am → 0700 hours	1.00 pm → 1300 hours
12.20 am → 0020 hours	11.45 am → 1145 hours	4.30 pm → 1630 hours
12.31 am → 0031 hours	11.59 am → 1159 hours	7.00 pm → 1900 hours
12.40 am → 0040 hours	12 noon → 1200 hours	10.10 pm → 2210 hours
12.52 am → 0052 hours	12.15 pm → 1215 hours	11.45 pm → 2345 hours
12.59 am → 0059 hours	12.59 pm → 1259 hours	11.59 pm → 2359 hours

CALCULATING AREAS

The letters 'a' and 'b' in the following formulae refer to opposite and parallel sides of the shape and the letters x and y are the diagonal lengths of a rhombus.

Area of a rectangle   = length × width
Area of a triangle   = ½ perpendicular height × base
Area of a parallelogram   = base × perpendicular height
Area of a rhombus   = ½ × x × y
(*Alternatively – as a rhombus is simply a special type of parallelogram – the parallelogram formula can be applied.*)
Area of a trapezium   = ½ × (a + b) × perpendicular height
(*A trapezium is called a trapezoid in the US and so some text books refer to trapezoid.*)
Volume of a cube   = length × length × length
Volume of a cuboid   = length × width × height
Volume of a pyramid   = ⅓ × area of base × height

USING PI

The ratio of a circle's circumference to its radius is called 'pi' and is usually represented by the Greek symbol 'π'. Pi is used in calculations involving circles, spheres and cylinders. Pi starts: 3.14159265358979323846264338327950288419716 9399... and then apparently continues forever with no known sequence or pattern.

Calculators usually have a button marked 'π' that displays pi to 7 or more decimal places (3.1415926). It is this pi that is most often used in calculations.

A handy way to remember pi to 7 decimal places is to remember the phrase: '*May I have a large container of coffee?*' The number of letters in each word represents the numbers in pi, starting with 'may' (3).

Here are some common mathematical formulae using pi (π):

Circumference of a circle
$$= 2 \times \pi \times r \ (2 \times \text{pi} \times \text{radius of circle})$$
Area of a circle $= \pi \times r^2 \ (\text{pi} \times \text{radius} \times \text{radius})$
Volume of a cylinder
$$= \pi \times r^2 \times h \ (\text{pi} \times \text{radius} \times \text{radius} \times \text{height of cylinder})$$
Volume of a cone $= \frac{1}{3} \times \pi \times r^2 \times h \ (\frac{1}{3} \times \text{pi} \times \text{radius} \times \text{radius} \times \text{height of cone})$
Volume of a sphere $= \frac{2}{3} \times \pi \times r^3 \ (\frac{2}{3} \times \text{pi} \times \text{radius} \times \text{radius} \times \text{radius})$

EASY REFERENCE TABLE FOR AGES AND STAGES IN NEW MONEY AND OLD!

	Preschool	} Foundation	3– 4 yrs
	Reception	} Stage	4– 5 yrs
Infants	Year 1	} Key Stage 1	5– 6 yrs
	Year 2		6– 7 yrs
Juniors	Year 3		7– 8 yrs
	Year 4	} Key Stage 2	8– 9 yrs
	Year 5		9–10 yrs
	Year 6		10–11 yrs
Secondary	Year 7 (1st year)	} Key Stage 3	11–12 yrs
	Year 8 (2nd year)		12–13 yrs
	Year 9 (3rd year)		13–14 yrs
	Year 10 (4th year)	} Key Stage 4	14–15 yrs
	Year 11 (5th year)		15–16 yrs
Sixth Form	Year 12 (lower 6th)	} Key Stage 5	16–17 yrs
	Year 13 (upper 6th)		17–18 yrs

ASSESSMENTS

Stage	Year	Assessment
Foundation Stage	Preschool	(sometimes known as Foundation 1)
	Reception	(Foundation 2)
Key Stage 1	Year 1	
	Year 2	At the end of Year 2, pupils will be assessed for Key Stage 1 SATs. These are not formal sit-down tests and the pupils should be blissfully unaware of any testing. An average-ability child will be expected to achieve a Level 2.*
Key Stage 2	Year 3	
	Year 4	
	Year 5	
	Year 6	At the end of Year 6, pupils will be assessed for Key Stage 2 SATs. These currently include formal tests in maths, English and science. An average-ability child will expect to achieve a Level 4.*
Key Stage 3	Year 7	
	Year 8	
	Year 9	At the end of Year 9, pupils will be assessed for KS 3 SATs. Formal tests have been replaced by Teacher Assessment. An average-ability child will expect to achieve a Level 5.* (Well 5½, to be exact.)
Key Stage 4	Year 10	
	Year 11	At the end of Year 11, pupils will do their GCSEs or equivalent.

* An average-ability child is expected to progress a level every two years. (Yes! Two years to progress one level.)

RESOURCES

Primary National Strategy, 2006
Primary framework for mathematics

The National Numeracy Strategy, 1999
Framework for teaching mathematics from Reception to Year 6

Key Stage 3 National Strategy, 2001
Framework for teaching mathematics: Years 7, 8 and 9

Key Stage 3 National Strategy, 2003
Targeting level 4 in Year 7: mathematics

Making Mathematics Count, 2004
The Report of Professor Adrian Smith's Inquiry into Post-14 Mathematics Education

Centre for Innovation in Mathematics Teaching (CIMT)
www.cimt.plymouth.ac.uk

Maths Enhancement Programme (MEP) Primary Demonstration Project

www.bbc.co.uk/schools/ks2bitesize/maths

www.bbc.co.uk/schools/ks3bitesize/maths

www.nrich.maths.org

INDEX